SCIENCE AS A **GATEWAY** TO **UNDERSTANDING**

International Workshop Proceedings
Tehran, '

Glenn Schweitzer and Yousef Sobouti, *Editors*

Office for Central Europe and Eurasia
Development, Security, and Cooperation
Policy and Global Affairs

NATIONAL RESEARCH COUNCIL
OF THE NATIONAL ACADEMIES

THE NATIONAL ACADEMIES PRESS
Washington, D.C.
www.nap.edu

THE NATIONAL ACADEMIES PRESS 500 Fifth Street, N.W. Washington, DC 20001

NOTICE: The project that is the subject of this report was approved by the Governing Board of the National Research Council, whose members are drawn from the councils of the National Academy of Sciences, the National Academy of Engineering, and the Institute of Medicine. The members of the committee responsible for the report were chosen for their special competences and with regard for appropriate balance.

Any opinions, findings, conclusions, or recommendations expressed in this publication are those of the author(s) and do not necessarily reflect the views of the organizations or agencies that provided support for the project.

International Standard Book Number-13: 978-0-309-12879-7
International Standard Book Number-10: 0-309-12879-X

A limited number of copies are available from the Office for Central Europe and Eurasia, The National Academy of Sciences, 500 Fifth Street, N.W., Washington, DC 20001; (202) 334-2376.

Additional copies of this report are available from the National Academies Press, 500 Fifth Street, N.W., Lockbox 285, Washington, DC 20055; (800) 624-6242 or (202) 334-3313; Internet, http://www.nap.edu.

Printed in the United States of America.

THE NATIONAL ACADEMIES

Advisers to the Nation on Science, Engineering, and Medicine

The **National Academy of Sciences** is a private, nonprofit, self-perpetuating society of distinguished scholars engaged in scientific and engineering research, dedicated to the furtherance of science and technology and to their use for the general welfare. Upon the authority of the charter granted to it by the Congress in 1863, the Academy has a mandate that requires it to advise the federal government on scientific and technical matters. Dr. Ralph J. Cicerone is president of the National Academy of Sciences.

The **National Academy of Engineering** was established in 1964, under the charter of the National Academy of Sciences, as a parallel organization of outstanding engineers. It is autonomous in its administration and in the selection of its members, sharing with the National Academy of Sciences the responsibility for advising the federal government. The National Academy of Engineering also sponsors engineering programs aimed at meeting national needs, encourages education and research, and recognizes the superior achievements of engineers. Dr. Charles M. Vest is president of the National Academy of Engineering.

The **Institute of Medicine** was established in 1970 by the National Academy of Sciences to secure the services of eminent members of appropriate professions in the examination of policy matters pertaining to the health of the public. The Institute acts under the responsibility given to the National Academy of Sciences by its congressional charter to be an adviser to the federal government and, upon its own initiative, to identify issues of medical care, research, and education. Dr. Harvey V. Fineberg is president of the Institute of Medicine.

The **National Research Council** was organized by the National Academy of Sciences in 1916 to associate the broad community of science and technology with the Academy's purposes of furthering knowledge and advising the federal government. Functioning in accordance with general policies determined by the Academy, the Council has become the principal operating agency of both the National Academy of Sciences and the National Academy of Engineering in providing services to the government, the public, and the scientific and engineering communities. The Council is administered jointly by both Academies and the Institute of Medicine. Dr. Ralph J. Cicerone and Dr. Charles M. Vest are chair and vice chair, respectively, of the National Research Council.

www.national-academies.org

NATIONAL RESEARCH COUNCIL STAFF

Glenn E. Schweitzer
Program Director, National Academy of Sciences

Merc Fox
Senior Program Assistant, National Academy of Sciences

Elizabeth D. Keller
Mirzayan Science Policy Graduate Fellow, National Academy of Sciences

Preface

In October 2007, the U.S. National Academies and the Iranian Institute for Advanced Studies in Basic Science organized the first of a series of planned U.S.-Iranian workshops on the topic "Science as a Gateway to Understanding." Several other Iranian organizations also cosponsored the event. They included the Iranian Academy of Sciences, Sharif University of Technology, and the Center for the Great Islamic Encyclopedia, where the workshop was held.

This new workshop series is a component of the broader effort of the National Academies to support bilateral workshops and exchange visits in a variety of fields with a number of Iranian institutions that began in 2000. The Institute for Advanced Studies in Basic Science is an important partner in this regard. Also, this workshop was a significant aspect of a visit by representatives of the National Academy of Sciences and National Academy of Engineering to Iran to review past cooperative efforts and to develop future programs.

The speakers at the workshop represented a wide variety of disciplines in the natural and social sciences and quite different perspectives on the theme of the workshop. Former Iranian President Mohammad Khatami was the keynote speaker. All attendees participated in their personal capacities, and the documents that were developed prior to and during the workshop express their personal views.

This report includes papers that were presented at the workshop and summaries of the discussions that followed some of the presentations. At the conclusion of the workshop there was general agreement that the presentations on many aspects of science and scientific cooperation that have a bearing on mutual understanding were an important first step. Several participants underscored that the next workshop should emphasize how scientific cooperation can lead in concrete terms to improved understanding among both academic and political leaders from the two countries.

Acknowledgments

Professor Yousef Sobouti and his colleagues at the Institute for Advanced Studies in Basic Science played the primary role in organizing the workshop and in compiling the papers included in this document. Together with former President Mohammad Khatami, he turned a vague concept into a highly productive workshop. This contribution is gratefully acknowledged.

This volume has been reviewed in draft form by individuals chosen for their technical expertise, in accordance with procedures approved by the National Research Council's Report Review Committee. The purpose of this independent review is to provide candid and critical comments that will assist the institution in making its published report as sound as possible and to ensure that the report meets institutional standards for quality. The review comments and draft manuscript remain confidential to protect the integrity of the process.

We wish to thank the following individuals for their review of selected papers: Yousef Bozorgnia, University of California, Berkeley; George Bugliarello, Polytechnic University; John Enderby, Institute of Physics (Bristol); Wilhelmine Miller, The George Washington University; Yves Quéré, Ecole Polytechnique

(Paris-Palaiseau); Geraldine Richmond, University of Oregon; and Barbara Schaal, Washington University.

Although the reviewers listed above have provided constructive comments and suggestions, they were not asked to endorse the content of the individual papers. Responsibility for the final content of the papers rests with the individual authors.

Contents

APPENDIXES

Words of Welcome

MOHAMMAD KAZEM BOJNORDI
Center for the Great Islamic Encyclopedia

Distinguished participants and dear guests from abroad, we are grateful to you for accepting our invitation. We have gathered here today to declare that science is a gateway to dialogue, peace, and friendship. We have gathered together to praise and respect mankind's achievements and to announce that science can facilitate rational human relationships and offer high quality of life to people everywhere. Scientists are among those social elites who define the cultural essence of each nation's civilization, and their common language for explaining phenomena provides the way for understanding among nations. We should learn this language from scientists and consider them to be the conscience of the greater society.

In conclusion, I would once again like to extend my gratitude to all distinguished participants for accepting our invitation, and I hope that this workshop will be a starting point for other conferences that convey a message of peace and understanding. Peace be upon you all.

1

Where Does Science Go?

HONORABLE MOHAMMAD KHATAMI
Former President of Iran

At the outset, I would like to express my gratitude and extend a welcome to those distinguished participants who have traveled from the United States of America, Senegal, and France. I would also like to welcome my distinguished countrymen who are present today. Finally, I wish to thank the Iranian Academy of Sciences, the Institute for Advanced Studies in Basic Scienceheaded by Professor Sobouti, Sharif University of Technology, and the Center for the Great Encyclopedia of Islam. I hope that this gathering will begin to bring hearts closer together, to create understanding, and to bring about the peace and tranquility that today's world needs more than ever. I also hope that this meeting will mark the beginning for future meetings and conferences to discuss and develop mechanisms for solving the problems that we all face.

As the West was waking from the so-called Dark Ages and embarking upon the new world that was ushered in with the Renaissance in the fifteenth century and the Reformation in the sixteenth century, the tale of Johann Faustus was published in Germany. Faustus is the story of an alchemist who made a pact with the Devil to achieve all his worldly desires, but with the consequence of burning eternally in the fires of Hell.

Twenty years later, the prominent British dramatist Christopher Marlowe published *The Tragical History of Dr. Faustus*, and the story finally entered into the Western psyche in Johann Wolfgang von Goethe's closet drama, *Faust*. Let us consider the story of western science—and modern science in general—to be parallel to the story of Faustus.

Dr. Faustus, a scientist and alchemist, was deceived by the Devil, but he never sold his soul to Satan. He eventually experienced different stages of torture and learned moral lessons through his torment. The deceived man returned to the domain of the soul after a long and painful journey to finally find righteousness through service to others—so much so that the angels praised him and took his soul to heaven. The Devil, whose power was manifested in his cunning and magic, was defeated by the majesty and grace of God.

By the end of the nineteenth century, Friedrich Nietzsche spoke of a mad sage, stating "(we) have killed God." Mankind had turned its back on kindness, beauty, and the mystery of magnificence, and society would have to pay for it. At the time, Nietzsche imagined a superman who could save him from his will. Short decades later, a "superman" manifested as Adolph Hitler would treat the world and mankind in a way that his predecessors—Europe's imperial colonists—treated a great number of people around the world.

Hitler dominated the majestic soul of the German nation and massacred millions in the western world. The western world fought back with an astounding modern technology, the atomic bomb, which ended the Second World War. However, those bombs did not kill Hitler, but rather decimated thousands of helpless civilian victims of the war.

At the end of the Second World War—a Hot War—commenced the even more frightening Cold War, which continued until the fall of the Soviet superpower and then ushered in an unprecedented and terrifying situation for the world and mankind.

In my view, the world has turned into a group of "superpowers" that consider themselves to be uncontested masters of the rest of the world because of the eradication of their powerful rivals during the Cold War. They rely on the false pride of their eighteenth century ancestors and expect the entire world to submit to their willpower and interests. They use whatever device they wish against those who oppose them, even when such measures violate contemporary ethics, law, and human rights. These modern "superpowers" employ humiliation, oppression, and hopelessness to facilitate the violence, terror, and destruction that ensures the power of the mighty and weakens the frail.

I do not deny the miraculous achievements of science and technology. Indeed, the foundation of modern human life rests on these two expressions of humanity. However, it should not be forgotten that science and its achievements led to the belief that man and nature could and should be dominated through technology. This philosophy has resulted in the increasing destruction of the natural environment and has justified the cultural devastation of those peoples who have historically been repressed. Modern science and the miracles of technology increasingly widen the gap between those in authority and those who are disadvantaged in the same that way religion, philosophy, communication, and other inventions of man's creative mind have been misused to the advancement of the powerful. Local authorities and politicians are rarely involved in the creation of science or technology, yet they often exploit the products of innovation to their benefit.

The third millennium was expected to begin with the soothing and promising call of the "Dialogue of Cultures and Civilizations," which could have been a lesson for a peaceful and secure life. Instead, our collective bright new future began with terrible explosions in New York City and Washington, D.C. that were followed by an atmosphere of fear. Tight security has been implemented at the borders of powerful nations throughout the world; and wars have been waged in the disenfranchised areas that escape those strictures resulting in disastrous campaigns, occupation, oppression, and the constant humiliation of many nations and societies. This dualism has fueled the extremism, violence, and terrorism that continue to generate fear throughout the world. At every stage, science and technology have served the purpose of violence and devastation, yet there is still a firm ground for optimism.

The matter of security can be and should be a universal right afforded to all people of every nation. Security should be everywhere, or no place will be secure. Sustainable security will be realized, however, only in the light of justice.

Moreover, the tradition of violence, occupation, and imposition has caused such suffering and pain that many scholars, politicians, and citizens in nations of privilege throughout the world have reconsidered this paradigm, increasing the likelihood of understanding and dialogue. We should strive to create a new tradition that offers the benefits of science to all of mankind and to honor ethics and beauty. This is the right way to salvage science and humanity. It is a difficult task, and scholars everywhere should collectively tackle it with all their might.

DISCUSSION

Glenn Schweitzer: President Khatami, you have a great deal of experience with science that has caused problems throughout the world, and though you have underscored the misery caused by the misuse of science, you remain optimistic that science can be turned in the other direction as a force for good. In your opinion, how can the scientific community increase the positive applications of science while trying to reduce the bad applications of science?

Mohammad Khatami: I leave the answer to you distinguished and learned men and women to show us how to benefit from science without misusing it. I will, however, make some short remarks.

Science and technology are the expression of man himself. So if man is good, the fruit of his efforts will also be good. Further, we can create a world in which people are rewarded for being good by developing cultures of benevolence that are bound by ethics and that offer the products of science and technology to the benefit of all.

Unfortunately, public opinion is largely influenced by power-hungry and profit-seeking lobbies. It should be a priority of those who serve the interests of justice, ethics, and goodness to put an end to this self-serving practice by entering into a dialogue among civilizations and cultures. It is hard to achieve, but it is possible with the help of all good people.

Norman Neureiter: Thank you for your interesting remarks. You have talked about the negative side of science. What bothers so many of us today is that violence seems to be part and parcel of the rhetoric of Islam. How do you reconcile the violence associated with religious purpose, revenge, *jihad*, and so on, with the misuse of science in general? Should science be practiced in the context of religion?

Khatami: It is a good and timely question. Violence in society is bad enough, but when it is used as a weapon by zealots to impose their religious beliefs, then it is horrifying. Unfortunately, for more than a century we have been living with global violence including two World Wars, with the damages of the first one alone exceeding those of all wars since the beginning of time. Following the Great Wars was the Cold War, which destroyed the young and tender economies, cultures, and diplomatic ties that were just emerging in the modern world and which set the calendar of peace and prosperity back by untold decades. Our shared legacy of this violent technology is the ever-existing terrorism pervasive throughout the world, but in a much more dangerous form and on a much larger scale than ever before. Thus, the modern era has been plagued by violence and the kind of intense, impersonal aggression that would have been impossible and unthinkable for the generations preceding the twentieth century, and this plague has culminated in an explosion of fear, oppression, occupation, and terrorism.

I am saddened to see that some of my friends attribute such violence to this or that belief or ritual. The sources of present-day violence can be found wherever injustice thrives, in all those places that suffer the legacy of harsh violence, particularly where the powerful humiliate the weak.

The only connection between religion and violence is that faith is the last refuge of the downtrodden, who lash out when they can take no more abuse. However, Christ preached love and the Koran carries the message of love. It is not insignificant that every chapter of the Koran begins with, "In the Name of God, the Compassionate, the Merciful." For the true followers of Christ and Islam, violence is forbidden and compassion is revered. Violence is considered to be a product of evil, and those who commit violence misuse religion and faith for their own evil causes.

Violence is the direct consequence of a logic that states "whoever is not with us is against us; and whoever is not with us, we are allowed to deal with in whatever way we like." In today's world this logic is claimed by those who commit terrorism as well as by those who fight terrorism. It is imperative to abandon this logic if the world wants to eradicate violence and replace it with the benevolence and justice that is common to all great religions and should be enjoyed by all societies. Let us work together to free the world from violence and those who perpetrate violence.

Thomas Jordan: You have introduced a new idea with the statement, "Security should be everywhere, or no place will be secure. Sustainable security will only be realized in the light of justice." The idea of sustainable security is an interesting one, and it would seem to many of us that this would require cooperation among nations, and in particular cooperation between the Islamic Republic of Iran and the United States. How do you suggest we achieve the cooperative spirit that is necessary to create sustainable security?

Khatami: I think this problem is bigger than Iran and the United States. We are all parts of a broken system that exacerbates misunderstandings and injustice. The differences between cultures and countries are immense and growing. The underdeveloped rightfully wish to move forward and prosper and to be free and independent, but find it difficult to prosper under the existing double standard in the world and the influence of the powerful in world affairs. This creates further misunderstandings and widens the gaps between peoples. We all should make an effort to ensure that the whole of humanity manages material and technological resources with equality.

Let me say one thing that I have said to my western friends before. Look at your western world through the spectacles of an eastern man. I, the easterner, have not seen much of western science and technology, and even less of western democracy. Instead, I, the easterner, have seen western colonialism and exploitation, the West's installation of corrupt dictators, and western support of coups d'état. I, the easterner, have been humiliated by the "democratic" and "advanced" West.

Under these circumstances, it is natural for the easterner to develop feelings of pessimism about the West. This pessimism, initially directed toward the bad policies of policymakers, gradually spread and was embraced by the whole of eastern civilization and culture. It is this pessimism that has prevented us from getting close to each other and solving our mutual problems.

Western statesmen should deny the legacy of exploitation and colonialism, both in its traditional and contemporary forms. The eastern statesmen should embrace the lessons that can be learned from the western world. Of course, this dialogue must be conducted in an atmosphere of fairness and on equal ground. If such an atmosphere is realized, I don't think there will be any difficulty in negotiations between Iran and the United States. I hope that the dialogue between scientists and people of knowledge will serve as an example to the statesmen.

Yousef Sobouti: The chair would like to thank President Khatami and all those who contributed to this discussion. However, I would like to make a comment. Science has a universal character to it—it is understood by everyone, everywhere, and at all times. Non-science doesn't have such character. By non-science, I mean politics, economics, and many other things. And I do hope that in the continuation of the program we will explore this universal character in the hope of achieving a better ground for common understanding.

2

Scientists and Truth

REZA DAVARI ARDAKANI
President, Academy of Sciences of Iran

Distinguished guests and scholars, welcome to Iran and to this workshop to discuss issues that are of critical importance to all societies. The world relies on its scientists to communicate truthful information and to offer efficient solutions, and I hope that your discussions will have fruitful results. A brief tutorial on the historic relationship between knowledge and understanding is in order as we begin this dialogue.

At the dawn of mankind's interest in logic and reason, Plato professed the possibility of creating a scientific, intellectual republic. This Platonic ideal was not forgotten and was later renewed in the world of Islam by Abū Nasr Muhammad ibn al-Farakh al-Fārābi. But it was not expanded very much until Sir Francis Bacon wrote his *New Atlantis*. In this work, he described a city governed by scientists and devoid of happiness and joy; those who looked upon it called it a mechanical city. Still, it was a peaceful and tranquil city.

Although history is not one of tranquility, peace, and understanding, the *New Atlantis* remained steadfast as a city of hope and expectation during a turbulent time, but with enduring concerns about its destiny. For instance, Kant, who idealized the Republic as a state of reason and peace, knew that such a republic would lead nowhere unless wisdom ruled it, and he expressed his concerns that it would face great dangers in the absence of such wisdom.

Until the mid-twentieth century, science remained the haven and stable pillar of hope in modern society, but the agitations that the world of modernity experienced were not so powerful as to shake its huge structure. Even though events in Europe during the first four decades of the twentieth century dashed much of that hope and hopefulness, and though certain doubts and a sense of hopelessness grew among some scholars, the principle of placing hope in science remained firmly entrenched in society.

Modern science provided an objective point of view and numerous benefits to the overall quality of life, which parlayed into an optimistic trust in the future. However, the question of science as a tool for peace and understand was rarely posed in the history of science.

The advents of Stalinism in the USSR, National Socialism in Germany, and global engagement in World War II were not experiences that could be ignored. These events were signs of the emergence of another age. After the war, the perception that science fostered agreement among non-scientists, that scientists were in agreement with each other on scientific issues, and that their differences could be resolved easily had changed, and people began to realize that scientific agreement did not extend to other domains, including that of culture, beliefs, and politics.

Today, almost every society depends to some extent on technology, but societies cannot follow the scientific model to establish universal consensus. Up until now, along with hope about the contributions of science, there has also been hope that a new culture of understanding would disseminate worldwide and replace friction between cultures. We have seen that European and American philosophers and scholars have forgotten about embracing a unique culture. No longer do they believe that science will show the way to the future of the world. Even those who have taken the fall of the USSR as the sign of the victory of liberal democracy have not hidden their despair about the establishment of peace, consensus, and understanding in the world.

Now science is more advanced than ever, and its dissemination in the form of consumption technology has resulted in uniformity all over the world. But the differences, wars, and misunderstandings have increased. The constitution of modernity, namely the Bill of Human Rights, has been ignored worldwide. Under such circumstances, how is it possible to remain hopeful about science?

In my view, science exists in the domain of truth, and scientists should be its children and followers. They should be committed to the spirit of loyalty to truth, goodness, and beauty, and to advocate science before the ruling powers as much as possible. In this way, science may become the gateway to understanding again. If we lose this hope and ignore the valuable role that truth plays in life, we become hopeless; and if we accept that science should serve at the mercy of politics and be used as a means for imposing enmity and power, then we should know that man is doomed to danger. It is still possible to rely on truth and to be hopeful, and this hope should be preserved.

This workshop signifies that hope exists. I hope that this meeting and its discussions will be a small step toward understanding.

3

The Innovation Ecology

WILLIAM WULF
President Emeritus, U.S. National Academy of Engineering

The National Academy of Sciences, National Academy of Engineering, and Institute of Medicine report *Rising above the Gathering Storm* (Committee on Prospering in the Global Economy of the 21st Century, 2007) and other reports have made recommendations to improve the climate for innovation in the United States, and by inference, internationally. However, many components of an "ecology" of interacting laws, regulations, policies, and institutions are not mentioned in the report—the intellectual property system, a broad tax policy that encourages investment, and a culture that encourages risk taking, among others. Two problems with the components of this ecology are immediately apparent. First, many of the components were designed for technology that was developed in the past, but not for technology that is being developed now or will be developed in the future. Thus, they are not optimal for achieving their avowed intent. Second, society almost never completely rethinks the nurturing of technologies, but rather it attempts to adjust the ecology in relatively small ways to reflect current reality. The result is an ecology that is far less than ideal for spurring innovation, and in some cases may even be counter-productive.

I will not dwell on the details of the aforementioned report, but the essential message of *Rising above the Gathering Storm* is that America's children and grandchildren may not continue to enjoy the level of prosperity that has been taken for granted over the last several decades. Innovation has been the driver of our success, and if we want to continue to prosper, we need to nurture innovation. The report makes some recommendations about how to do that:

- First, we need a strong technical work force, and science and technology education are necessary to this endeavor.
- Second, we need to generate good ideas, and we should strongly support basic research.
- Third, we need to attract the best minds to science and engineering education.

In addition to these crucial components, a larger "ecology" that supports innovation, permits and even encourages risk-taking, and offers "patient capital" to the entrepreneur should be fostered. Laws and regulations must protect the public while simultaneously encouraging experimentation through tax laws that support and reward investment and intellectual property (IP) laws that adequately and appropriately protect intellectual property. The list of essential components in the innovation ecology—the collection of interacting and interdependent policies and activities that support innovation—is extensive. Perhaps most importantly, the world is no longer a set of independent economies, which means that each component must interact harmoniously with its foreign counterpart.

Let me illustrate the two points above, namely that the ecology we have today was invented for the technology of yesterday, and that the changes to the components of the innovation ecology have generally been incremental. I will give examples from the U.S. experience because I know that system best. I believe all these examples, however, have counterparts in other countries.

Let's start with the current patent system. The system was originally designed to protect large physical machines, which were based on the design of their physical models that were, until very recently, filed with the patent application. It seems unlikely that this system designed for large machines can be applied equally well to the protection of software, pharmaceuticals, snippets of DNA, or business processes—all of which are now covered by our patent system.

I spoke to a group of about 30 Chief Technology Officers from California in 2005, and each one of them said that the patent system was broken. They only continue to patent items for defensive reasons, so that they can trade patents with those firms that prefer to buy patents of developed products rather than to invest in competitive research. They said that from the point of view of promoting innovation, the current patent system was irrelevant and in some cases even counterproductive. Particularly in the information technology (IT) industry, the speed of the product cycle is so great that the notion of protecting technology for decades is irrelevant because the patent system is rooted in protecting large machines, not software or biotechnology. It is not tuned to technology as it exists today; and for them, the patent system simply is not working.

But it's even worse than that. Much innovation comes from small companies. However, these companies often cannot afford to build a portfolio of defensive patents, and they are disadvantaged relative to older, larger, and perhaps less innovative companies. The patent system was created to foster innovation, but it may be having exactly the opposite effect by disadvantaging small companies.

Clearly, intellectual property protection is important as a component of the innovation system, so let's address copyright. The very idea of a prohibition on "copying" as the mechanism of protecting some intellectual property does not make sense for many twenty-first century technologies.

Every once in a while I encounter a web page containing the copyright symbol. This page was copied at least a half dozen times on its way to my screen. It was copied from the hard drive of the server to the primary memory of the server. It was copied from the server's main memory onto the Internet. The page may have been copied anywhere from one time to several hundred times in the process of being transmitted over the network. Finally, it was copied from the Internet to my computer's primary memory in order to be displayed on my screen.

Of course, the person who put the copyright symbol on that web page didn't mean to prohibit *that* kind of copying, which is essential in displaying their page on my screen. It is perhaps ironic that the web page with the copyright symbol would have no value if it hadn't been copied. If it were just stuck on the hard drive of the server, it would be of no value whatsoever.

The problem is that this kind of copying is absolutely indistinguishable from the kind of copying that the author didn't want to happen. The difference has to do with the *intent* of making the copy—whether to display the page or to steal its content. At the level of the machine, there is no way to distinguish between the two.

The concept of prohibiting copying as a means of protecting intellectual property makes sense when artistic or literary expressions are on physical pieces of paper. It doesn't make sense when it comes to digital information. "Copying" is simply the wrong concept to use in the digital world as the tool to enforce what the copyright laws were intended to accomplish.

Let me address another aspect of intellectual property—the activities of the Food and Drug Administration (FDA). A member of the National Academy of Engineering who is a very successful entrepreneur has been involved primarily with medical device companies, although recently he has branched out to deal with some treatment techniques as well. He claims to have an extremely effective cancer vaccine that has been given to a person diagnosed with cancer. This vaccine destroys the cancer by creating an immune system response to it, and he says that it is extremely effective against tumors that are otherwise very difficult to treat.

What is the problem? Well, it is unlikely that this vaccine will ever be sold in the United States because the FDA test of safety and efficacy requires a randomized, double-blind clinical trail. The problem is that this vaccine was manufactured for *one* specific patient and *one* specific tumor; it is the ultimate product of personalized medicine. Since there is only one tumor in a specific person against which it is believed to be effective, it is impossible to do a randomized, double-blind test. Thus, under current FDA policy this therapy cannot be approved.

Isn't it ironic? A procedure created to ensure safety and efficacy is, in fact, preventing access to what my friend claims is effective therapy. Once again, a procedure created for an old type of technology doesn't work for a new type of technology.

Let me move on to the antitrust laws which are a component of the innovation ecology and which help make room for new entrants into a field. These laws were passed in the late nineteenth century, in the era of railroads and steel monopolies. They were developed in the context of the economic theories of the time that equated value with scarcity. Diamonds were more valuable than gold because diamonds were scarcer; gold was more valuable than copper because gold was scarcer, and so on. If we had to compete in the marketplace to buy diamonds because they were rare, we would pay a relatively high price for them.

However, with software and other IT market items, the premise of scarcity as a determinant of value is backward. Value is not related to scarcity. Instead, it is related to ubiquity. You may have heard of Moore's law (the number of transistors per unit area doubles every 18 months), but you probably do not know about Metcalfe's law. Robert Metcalfe invented the Ethernet, and his law says that the value of a computer network is proportional to the square of the number of nodes connected to it. If only one person in the world owns a telephone, it's not very valuable. Its value goes up as more and more people have telephones, and it is at maximum value when everyone has one.

The same principle holds true when it comes to certain software. For example, I use Microsoft Word. I do not use it because it's the best word processor, or because it is the most error-free, or because it is the cheapest. I use it because I am reasonably confident that if I create a ".doc" file and attach it to an e-mail and send it to you, you'll be able to open it, edit it, and send it back to me. It is the ubiquity of Word that makes it valuable to me, not its scarcity.

A law based on the economics of scarcity is not likely to be effective in an economy of ubiquity. Moreover, the remedies in the antitrust system do not reflect the economics of ubiquity, either. Take the company Microsoft as an example. Microsoft has a monopoly, but a few years ago the courts supported Microsoft when the government tried to enforce antitrust laws.

I think that the judge made the right decision given the context of the case. That is to say, it would not have made any difference whatsoever if Microsoft had been broken up into a company for Windows, a company for Word, and a company for Excel. The motivation to use Word would have remained exactly the same.

Now let's address the problem of tax credits for company expenditures on research and development (R&D) and the U.S. Congress's annual authorization of those tax credits. The same set of thirty chief technology officers that I mentioned above agreed that the R&D tax credit had no influence whatsoever on their R&D investments.

The reason is quite obvious: R&D takes many years. If they invested this year to take advantage of the R&D tax credit and then next year there was no tax credit, they may have to stop their research; and they would have just wasted money that they had spent. It is only with the assurance that such a credit will last for a reasonable period of time that it can have a meaningful effect. That doesn't necessarily mean it should be made permanent. It just means that it must be valid over a reasonable number of years.

Each of these aspects that I have addressed is an example of how our current innovation ecology was designed for the past and not the future. I have one final example which concerns the export of technology.

The intent of export controls is completely logical; it prevents dangerous exports, such as military aircraft to terrorist groups, for example. On the other hand, the implementation of U.S. export controls is broken and counterproductive when we control items that are widely available the world over from non-U.S. suppliers. Thus, in many cases the controls only have the effect of damaging U.S. business.

The pace of technological change is accelerating. Thus, even if we managed to fix every one of the components of this innovation ecology to be just right for today and tomorrow, they probably wouldn't be right for the day after tomorrow. Accordingly, our solution must involve a process by which we are able to periodically stand back and evaluate the intent and methods of implementation for intellectual property protection, import/export control, and antitrust laws. In other words, what we need is not just a set of changes to patents, copyrights, and so on, but an institutionalized process for renewal that is also relevant to the formation of laws and regulations.

DISCUSSION

Yousef Sobouti: My impression has been that scientific discoveries and technologies have developed almost in parallel to each other. Do you maintain that in the past few decades, scientific discoveries have taken a leap forward and left the technologies behind?

William Wulf: No, I wouldn't say that. There is a wonderful book entitled *Pasteur's Quadrant* (Stokes, 1997). It characterizes scientific research along two axes. One axis represents the pure search for understanding and knowledge, yes or no. The other axis represents the desire to solve a practical problem, yes or no. The quadrant which expresses neither interest in understanding nor interest in practical application is the quadrant that we don't need to worry about. The usual description of pure basic science is the quadrant that defines the search for knowledge without respect to application. The usual description of applied research is the quadrant that defines the desire for application but not necessarily for basic understanding.

But there is an important fourth quadrant that seeks basic knowledge and has an application in mind as well. That is "Pasteur's Quadrant," and it is not pursued as vigorously as it should be. I am not particularly worried about the purely applied and not particularly concerned about the purely basic science. Of course, both could use more money. I believe that this is true in Iran, and I am sure that this is true in the United States. But perhaps we haven't given enough emphasis, prestige, or value to that quadrant of both of our concerns.

Hydari Khajehpour: Most of the activities which you have described are more or less related to the private aspect of innovation, including the property rights of innovators and the legal aspects of those rights. I would mention a point that has been almost overlooked in the United States, in other western countries, and nowadays in developing countries such as India and Iran; namely, public compensation for invention and discovery. Many innovations are not the product of a single person's activity or a few persons' activities in a laboratory; instead, they have been developed based on the cultural knowledge that has no legal protection. Many drugs or methods of treatment are examples. Such drugs or therapies are picked up by pharmaceutical companies and patented after further development. How are the rights of the culture protected in these cases?

Cultural property rights all over the world are almost totally ignored. Many cultures' knowledge has been developed for profit, but their societies have not been compensated in any way. Many local drugs have enjoyed centuries of prior experimentation recorded in the public record of history, but now are patented as and protected as private property. Related to these public property rights is the investment spent by governments and by public institutions. Legal aspects of public property rights should also be taken into account. Is this also a part of your suggestions?

Wulf: That is a very interesting and useful insight. I think it is correct. There was recognition in the United States beginning in the late 1980s that there was a backlog of scientific discoveries, principally within universities, that were not being used to benefit the public. This knowledge had been paid for by the public. Well-intentioned legislation was passed by the U.S. Congress to give such property rights to universities so that they would have an incentive to develop discoveries for the greater good. However, in my estimation, the approach didn't work and we need another idea. But you are right. There is a huge amount of knowledge that was paid for by the public that only slowly filters back into benefits to the public.

Etienne Guyon: I would like to come back to your comments on Pasteur's Quadrant. When Louis Pasteur introduced his vaccine against rabies, he gave shots to a child named Joseph Meister without going through the usual preparatory steps before administering shots. In other words, Pasteur did something completely inappropriate that he should never have done—even in his time it was recognized as an incorrect technique. In France, as in other countries, we promote a principle of precaution which precludes the practice of an activity if there is a slight risk that the activity will be harmful. How can we reconcile this principle of precaution and its impediments with the approach of a man like Pasteur, whose disregard for correct aseptic technique resulted in the first effective treatment for rabies? Is there a solution, a half-way solution?

Wulf: The precautionary principle is that we should not do something unless we can be absolutely certain that there will be no negative effects. Fairly recently, our Environmental Protection Agency together with the Food and Drug Administration enacted a ban on silver nanoparticles for use in antibiotics until it is demonstrated through experimentation that they have no negative effects on the environment or on people. Of course, anyone who knows mathematics understands that it is impossible to prove the negative.

The general public does not understand this risk or even the notion of risk. I think one thing that we need to do is to educate our public about the notion of risk. The concept that something can be totally without risk is just fallacious. Understanding the need to be very careful, we cannot guarantee a complete lack of negative effects in any circumstances. It is just impossible.

REFERENCES

National Academy of Sciences, National Academy of Engineering, and Institute of Medicine. 2007. Rising Above the Gathering Storm: Energizing and Employing America for a Brighter Economic Future. Washington, D.C.: The National Academies Press.

Stokes, D. E. 1997. Pasteur's Quadrant: Basic Science and Technological Innovation. Washington, D.C.: Brookings Institution Press.

4

Wisdom: the Best Gateway to Understanding

MOSTAFA MOHAGHEH DAMAD
Academy of Sciences of Iran

I would like to begin my speech with a comparison between the words *science* and *wisdom*. The definition of *science* is generally clear to everyone, but the meaning of *wisdom* seems to be more complex, so I have chosen this term to mean the equivalent of the Arabic *hikma*, but I am not sure whether it is used in the same sense in the western world.

Hikma appears more than 20 times in the Holy Koran, where it is of eminent significance. It is repeatedly emphasized that the Prophet Muhammad teaches people "the Book and the Wisdom," and on one occasion it is said, "He [God] grants wisdom to whom He pleases, and to whomever is granted wisdom, He indeed is given a great good.". Moreover—and this is probably more important—in the Koran, God Himself has been called "wise" (*hakim*) more than a hundred times. In this context I interpret the concept of *hikma* to be "intellectuality" joined with spirituality, education, cultivation, and ethics.

Now, science has certainly been capable of bringing ease and prosperity to mankind. But since mankind is composed of a human and an animal dimension, it cannot be doubted that science, in turn, can become a means to satisfy man's arrogance, to generate his vices, to sow differences, and to spread conflicts, wars, and bloodshed.

In Islamic literature, science has been divided into two groups: useful and harmful. The former leads to the development and promotion of human society, while the latter causes corruption on Earth and the destruction of humans and nature. Science is useful as long as it makes mankind humble toward the magnificence of being, and useful science goes hand in hand with spirituality and ethics. On the other hand, harmful science engenders man's arrogance to a degree that he neither shows any regard for his natural environment nor for his fellow man.

It is because of the aforementioned scientific arrogance that modern man finds no reconciliation with the earth and the very environment in which he lives. On the contrary, he pollutes and destroys it ever more; man's arrogance is a result of his excessiveness in the sciences.

The excessive belief in the sciences, or put differently, a dry and petrified scientific attitude devoid of any spiritual background, an interpretation of the world from the narrow angle of the natural sciences, is the main reason for the destruction and corruption of mankind's natural environment. This is the most conspicuous result of the industrial developments of the last few centuries. Science has brought forth man's unlimited arrogance. In modern culture, science has replaced religious belief and thus has become man's true master.

It was Auguste Comte, the science-praising French philosopher, who first divided the development of our knowledge into three historical phases: first, the divine or godly phase; second, the philosophical or abstract phase; and third, the scientific phase (Comte, 1853). In the divine phase, man attributes all phenomena and all changes in the world to God's will or to supernatural forces. In the philosophical phase, man's intellect becomes capable of experimentation and the art of abstraction, thus relating the development of natural phenomena to forces that are themselves hidden, but whose effects are apparent. In this stage, man searches for a subjective and final explanation for the natural world. In the last phase, which is scientific and experimental, both man's reason *and* man's vision become subject to observation and trial so that only the palpable and the visible are valid.

Comte held that mankind has left behind the first two stages and has reached the third one. Man no longer quarrels about things that bring no benefit to him; rather, he leaves aside useless argumentation and turns exclusively to whatever is beneficial to his actual life.

Comte thought that religion would not be accepted or followed if it was not accepted by men of learning. In addition, he said that modern science can only accept and worship one single being: the being of humanity that stands above everything else and any other being, encompassing all past and future generations. The being that must be worshipped is comprised of all those who have endeavored for the progress of mankind and who will continue this way in future times. Toward the end of his life, he had enjoyed the taste of love, and on the basis of his philosophical convictions he founded the *Religion de l'Humanité*and he built a temple to hold prayers.[1] Comte identified humanity as *le Grand Etre* and he called himself *le Grand Pretre,* noting of course that in his religion, *worship* did not have the usual meaning of glorifying, but of caring and attending (Foroughi, 1989).

Comte thus stated emphatically that science would be mankind's future religion. We see that his prediction was not too far afield from reality, because in recent centuries the sciences have indeed become man's ultimate idol—a deity to be worshiped by all people. It has become even more than a deity, namely a monopolistic god that tolerates no rival or partner—an idol devoid of life and soul that has slaughtered without trial any concept that was filled with meaning and soul. Spirituality, ethics, philosophy—be it natural or metaphysical philosophy—has no status without the validation stamp of the sciences. Today, scholars have left behind the divine and metaphysical stages, and any religion that they can accept and believe in must adapt itself to experimental science. In other words, science is mankind's future religion.

---

[1] The spread of Manaichaeism through Spain and into neighboring countries probably had a strong influence on Comte.

During recent centuries, humanity has turned ever more to the sciences in order to gain an easier and better life. But the kind of science that interpreted man's environment as separate from the warmth of life and devoid of any spirit and meaning has made him forget his inner values and the spiritual nature of the world, and has led him toward a continuously narrower and more painful life. Man view of nature has narrowed to such an extent that he has literally made breathing difficult for himself. Science, which had set out to help and to attend mankind, became its deadly enemy, quite according to what Sheik Saadi, an outstanding Iranian poet of the thirteenth century said: "A servant went out to fetch some water, but the water rose and swept the servant away."

Present-day science has made nature a tool and a meaningless object at the hand of modern city dwellers. It has secularized life, having separated it from divine grace. Nature is no longer a mirror reflecting God's beauty. Not only has nature lost its connection with modern man, but mankind also finds itself separated from nature, which has become a stranger to man, void of any holiness. If anything is holy to modern man, it is nothing but his own self; nature is only a reservoir to be exploited for man's needs and joys. Hence, modern man does not have a friendly eye on nature but looks upon it only in a materialistic, exploitative, and selfish way.

Nature is no longer man's beloved one to whom he could show his love, not his partner for whom he might feel responsible while at the same time enjoy its company. Instead, man looks upon nature as a woman of pleasure who can be used for his bodily pleasures without any sense of responsibility. Such a prostituted nature, as a result, has gradually been worn out and become hardly useful anymore, as though having reached its final days, lying old and dull before man who has no use for it anymore.

Of course, man has always had an innate propensity to harm and dominate nature and to use it for his needs and pleasures. But it was modern science that prepared a theoretical foundation for such an attitude toward nature by desecrating it. Modern man finds no spiritual concept anymore for which to reach out in the heights of the mountains, in the depth of the oceans, or in the endless skies. On the contrary, it seems that their sublime magnificence annoyed man's dominant and presumptuous spirit, provoking him to deprive them of their natural magnificence by controlling them. No longer does modern man aspire to the spiritual experience of ascending to the divine kingdom, as we find in the story of Christ, in Dante's *Divine Comedy*, or in the Prophet Muhammad's nightly ascension to the skies. Man has become arrogant by ascending mountains, flying spaceships, and wandering between the planets. Now man dances on the ruins of what he has left behind. In its desecration of nature, science has been successful to such an extent that even many pious people have lost their religious feelings toward nature.

In short, science devoid of spirituality might enhance mankind's power, but because man applies the philosophy of instrumentalism toward nature in his drive to dominate it, he has become so arrogant that he has trampled almost all ethical considerations. On the other hand, *hikma* declares the world to be a comprehensive mirror of God Almighty, which reconciles man with his natural environment so that man can walk along the path of salvation. *Hikma* views the world as holy, whereas science tries to remove all holiness from being.

DISCUSSION

Mostafa Damad: I am sure that there are many questions, for all of the people in the audience are scientists, and I have criticized science. I should remind everyone that I am a professor of Islamic studies, and I should defend religion and make peace between religion and science. I think that in the word *hikma*, there is peace. The word is used time and again in the Koran. I tried to find this word in Hebrew, perhaps in the Old and New Testament. In particular I was interested to find the word that is opposite in meaning to *Hikma* in the Old Testament, but I failed.

Masoumi Hamedani: In most English texts translated from Arabic, the word *hikma* is translated as "wisdom" and sometimes as "theosophy."

Damad: But in Koranic texts *hikma* is not synonymous to philosophy or theosophy.

William Colgalzier: I like your title, "Wisdom: the Best Gateway to Understanding." However, it is difficult for me to accept blaming science for the inappropriate application of science. We use science as a way to understand the natural world through the use of experiments. Science is not a substitute for religion, and it's not a substitute for human values. Certainly it's not a substitute for wisdom, which is often in short supply. My view is that science is a tool. In looking at the application of science, hopefully it's a tool that is used wisely. It could certainly be used badly as well.

Damad: Professor, you are a religious man and a scientist. But I know many other scientists who say that science does not have a religion or philosophy, and that they would believe in God if only they could find him in the laboratory. You are not this type of scientist. You perhaps think like me and would shake hands with me.

Mehdi Bahadori: Why don't you use the word *ethics* instead of *wisdom*?

Damad: Wisdom encompasses ethics. Wisdom has a very vast meaning. It is a combination of ethics, knowledge, spirituality, education, etc. All of these concepts are included in *hikma*, according to Allameh Tabataba'i and other Islamic philosophers (Tabataba'i, 2003).

Bahadori: Is the word *ethics* used in the Koran?

Damad: Yes. Where God addresses the holy prophet, "you are endowed with *khologhen azeem*," i.e., with "great ethics."

Yousef Sobouti: I would like to follow the same theme as Professor Colglazier. A newborn baby does not have wisdom. It has the potential to become wise, but it does not possess wisdom when it is born. Wisdom comes through experience and learning at later stages of life, but it is science that has brought the largest and vastest understanding of nature to man.

Damad: What are you trying to conclude?

Sobouti: I am trying to agree with you. Wisdom could be a gateway to understanding, but wisdom comes through experience in life, and the most important means of gaining this experience nowadays is science. The examples that you gave from Auguste Comte or from the earlier philosophers belong to the past. The points of view of people have also changed. I don't think Auguste Comte's teaching would be acceptable to many scientists today.

Damad: Professor Sobouti, let me answer you by asking a question. Do you think if I were a scientist, but without the rationality that you advocate, or the ethics that Professor Bahadori emphasizes, or the wisdom that I have described, it would be possible for us to be at peace together and live kindly and lovingly? Do you really think that this would be possible?

Sobouti: That's a difficult question. I am not claiming that I have a prescription for solving the disputes of the world. But what I'm saying is this: the concepts that you use—ethics, for example— are manmade values. They depend on the elements of culture; on habits, on practices, on religions, and on many other things. Cultural values cannot solve disputes. On the other hand, science is free of such values. It doesn't depend on any culture, on any belief, on any geographic location, or on any historical era.

Damad: I disagree. I maintain that ethics is independent of manmade values. If we live in peace together, it is because of ethics. With science only, we will have wars and conflicts with each other.

REFERENCES

Comte, A. and H. Martineau. 1853. The Positive Philosophy of Auguste Comte. London: John Chapman.

Foroughi, M. A. 1989. The development of philosophy in Europe. Tehran: Safi Ali Shah Publication.

Tabataba'i, M. H. and A. Q. Qarai. 2003. The Elements of Islamic Metaphysics. London: Islamic College for Advanced Studies.

5

Interacademy Cooperation: An Approach to Understanding

ABULHASSAN VAFAI
Sharif University of Technology

While the basic needs of science have not altered in the last century, social demands have changed dramatically. Science must address complex issues that are global in scale, and it must deal with difficult problems that can only be solved through international collaborative efforts. In its essence, science is a collective enterprise—advances owe more to the pain-staking dedication of teams rather than the efforts of individuals. Cooperation is at the heart of scientific endeavor and dialogue, and they in turn are vital mechanisms for collaboration and building reciprocal interests as well as mutual understanding among people of different nations.

It has been my privilege to host and participate in a number of gatherings regarding scientific cooperation. I remember with satisfaction the first steps taken toward international scientific exchange between members of the American and Iranian Academies of Science. Having paid visits to our respective countries and having participated in discussions with various representatives of the relevant institutions, a number of promising proposals were made and put into action with unprecedented success.

The results of our first endeavors were particularly encouraging and established the precedence for almost a dozen activities such as workshops and panels in various areas of engineering, science, and technology. These topics incorporated research into fields of great significance to the people of our country, such as water, earthquake science, nutrition, health, and environment.

I would like to briefly summarize the major cooperative activities that have taken place during the period of 1999-2007:

WORKSHOPS

- Experiences and Challenges of Science and Ethics, 2002, Bellagio, Italy
- Ecology of the Caspian Sea, 2002, Tehran, Iran
- Higher Education, 2002, Tehran, Iran
- Water Conservation, Reuse, and Recycling, 2002, Tunis, Tunisia
- Food Safety and Food-Borne Disease Surveillance Systems 2004, Tehran, Iran
- Drought Forecasting and Management, 2005, Tehran, Iran
- Science and Society, 2006, Toutour, France

OTHER EVENTS

The 2007 visit of a delegation from the U.S. National Academies included a workshop at Sharif University of Technology (SUT) at which scientists discussed bilateral scientific relationships in fields of interest to both parties. One of the highlights of that workshop was a lecture given by 1993 Nobel Laureate Joseph Taylor, attended by more than 1,000 young scientists from all over the nation. It was a motivating force and a source of inspiration for all young students in the field of astrophysics. After serious deliberation and discussion, several items were accepted for establishing the following projects:

- a workshop on earthquake engineering to discuss unreinforced masonry structures (early 2008)
- an exchange of students and professors in the fields of information technology and physics, with institutions of higher learning to be finalized later
- a visit by Professor Thomas Schelling, a Nobel Laureate in economics at the University of Maryland, in late December 2007 to SUT to give a lecture and to discuss possible bilateral scientific relationship
- a visit by Professor Burton Richter, 1976 Nobel Laureate in Physics, in May of 2008 to SUT

All of these activities are the result of the collaboration of scientific minds in the pursuit of cooperation and understanding amongst nations. These original meetings paved the way toward the fulfillment of an ambition to gather together the best minds from around the world without political impediment and to incorporate their humanitarian ideals and scientific knowledge into action for the betterment of humankind.

Bilateral scientific cooperation between SUT and the University of California, Davis has also developed. Five Iranian students from the Graduate School of Management and Economics at SUT are now in the United States.

These are just a few of many projects in recent years. It is through the dedication of people within scientific institutions that such models of bilateral scientific cooperation can be implemented.

Although there has been a most fruitful beginning, even more can be attained by positive allegiance to the spirit of international scientific dialogue and a confirmation of our commitment of the promotion of peace and open scientific exchange.

Our meeting here today is a reiteration of our belief in the building of a prosperous and equitable world through the expression of international scientific cooperation and collaborative action. I am optimistic that with interacademy cooperation and the support of the leaders and members of our two establishments, the future holds further steps forward from the original milestone laid by this pioneering agenda, and the essence of mutually beneficial scientific cooperation can engender an even better future.

I trust that this workshop will be the beginning of a very successful endeavor toward achieving our goals of understanding through science.

Of one Essence is the Human Race
Thusly has Creation put the Base,
One Limb impacted is sufficient,
For all Others to feel the Mace.[1]

---

1 This poem by Saadi Shirazi calls for breaking all barriers between peoples and graces the entrance to the Hall of Nations of the United Nations building in New York.

6

Understanding Others, the Science Way

YOUSEF SOBOUTI
Institute for Advanced Studies in Basic Science

In their study of nature, physicists often resort to the reduction to basics. However, most natural phenomena are complex; and when confronted with such complexity, a physicist tries to identify the prominent features of the phenomenon, strip away the insignificant details, and reduce the problem to a manageable and understandable model. As a student of physics, I wish to follow the same procedure to understand others.

Astronomy as a study of the skies became an exact science as early as the times of Hipparchus and Ptolemy. Through observations of the motions of the heavenly bodies, inquisitive men had understood the order prevailing in the skies and were able to predict astronomical events, such as tides, eclipses, conjunction, and opposition with incredible accuracy. Similarly, ancient geometry, born out of everyday practices in land surveying and building construction, also became an axiomatic science at about the same time.

No one disputed the legitimacy of these two disciplines. They were appreciated by everyone, irrespective of social and cultural status. They could be taught and learned in any language and by anyone who was interested. At no time or place did their tenets become sanctified, nor were any of their practitioners promoted to the state of sainthood. In short, astronomy and geometry emerged as two culture-free intellectual constructions of man's mind as early as 20 centuries ago.

All this was possible because both disciplines were observation-based and relied on natural facts to support their conclusions. These facts left no room for dispute, or rather they offered a built-in mechanism to resolve disputes. One could convince or be convinced by one's fellow practitioners through logical reasoning and turn to the facts as the supreme arbitrator. In what follows, we expand on this culture-free and dispute-free nature of some of the contemporary sciences in the hope of turning away from controversy and toward "understanding others, the science way."

Unlike astronomy and mathematics, other creations of man's intellect were not so blessed. Physics, the modern terminology for the invisible sciences of the ancients, had to wait until the era of Galileo and Newton in the sixteenth and seventeenth centuries, respectively, to begin its axiomatization, which still is being revised and refined. In spite of their astonishing achievements, chemistry and biology are still in their infancy, and the social and psychological sciences have at best emerged as empirical disciplines. Supernatural ideas and beliefs are not represented by any formal scientific discipline. Why has it taken so long for most of the natural sciences to arrive at acceptable levels of clarity and to be perceived and understood as value-free?

One astonishing and almost universal tendency of the ancient thinkers was their holistic approach to the observation of nature. In contrast, the practice of modern science divides complex issues into small components in an effort to understand them stepwise, from the simple to the difficult. A consequence of the ancients' lofty and unachievable goal was the tendency to resort to metaphysical concepts whenever hypotheses fell short of factual evidence. Ad hoc as they are, such notions differ from time to time, place to place, mind to mind, and culture to culture. As such, they potentially nurture the seeds of controversy. Then, in order to defend them, when confronted with opposing viewpoints, man invariably has looked for support from believers, patrons, and patron-institutions. Let us look at some historical examples.

In the fifth century B.C., Socrates was tried by a jury of 500 Athenian elites. The main charges against him involved the divergence of his philosophical points of view from the accepted values of Athenian society. In 400 B.C., both the philosophical ideas of Socrates and the social values of Athenian society were vague concepts. Neither the defendant nor the prosecutor was able to provide unequivocal evidence to support or discredit claims and counterclaims. The result was tragic: Socrates was convicted and made to take a deadly potion.

Centuries later, a bigger tragedy took place. The teachings of Jesus of Nazareth and those of the orthodox faith of his community confronted each other. Both sides were committed to their doctrines and had disciples and believers to defend their causes. The logic of one side, however, was not acceptable to the other. Inevitably, it ended in tragedy.

Throughout history, such tragic episodes have repeated themselves. The pattern is always the same: two factions oppose each other over a vaguely conceived cause, such as a religious belief, a social value, a moral code of conduct, a philosophical doctrine, or a material interest. The opponents differ in their logic, the disputes remain unresolved, and unjustified measures are used.

Let us consider examples from Muslim society in the first millennium. Abu Nasr Farabi (873-949) and Abu Ali Sina of Avicenna (980-1037) were undoubtedly the greatest philosophers of their times, as well as devout Muslims. Abu Hamed Ghazzali (1058-1111), an equally renowned thinker and a great theologian, however, was at odds with these philosophers. He maintained that the teachings of philosophers, including mathematics, weakened the pillars of the faith. He called Farabi, Abu Ali and, for that matter, all philosophers heretics. Fortunately, the Islamic societies in their flourishing period between the seventh to the twelfth century were tolerant enough to let the verdict pass without harsh retribution. Ghazzali's defiance of philosophy and intellectual reasoning did, however, leave long-lasting impressions for years to come. The great theologian had zealous followers amongst the elite and the commoners, and their influence eventually led to the suppression of free thought and the acceleration of intellectual decline within Islamic societies.

Let us proceed to sixteenth- and seventeenth-century Europe. The Ptolemaic model of the geocentric universe combined with the Aristotelian viewpoint that man stands second to the Almighty in honor put the earth in a noble position in the scheme of creation. Somehow, this notion worked its way into the teachings of the Church. Taking the earth out of the center of creation was a sacrilegious act; and Copernicus, fearing his fellow theologians, chose to postpone the publication of his heliocentric theory of the universe to the very last day of his life in 1543. Galileo (1564-1642) was wise enough to deny altogether the motion of the earth in the Court of Inquisition and avoid any unpleasant consequences.

In the early twenty-first century, many of the natural, human, and social sciences have achieved acceptable levels of universal clarity, and their practitioners have learned to reconcile differences through sober dialogues. This is a welcome development, yet there are many global issues that are not satisfactorily cast in objective terms and other issues that may never be viewed objectively. The following are examples:

- Economics: Is it better to allow open competition in a free market at the risk of wiping out the disadvantaged who lack the resources necessary to compete, or is it better to allow the state to control the production and consumption of goods at the risk of corruption?
- Governance: Should a democratically-elected ruling body answer only to its own electorate, or should it be accountable to its neighbors as well?
- Human rights: Are human rights defined by western ethics or eastern standards?
- Ethics: Are they consistent throughout the faithful Christians, the devout Muslims, the Jews, and the believers in other faiths?
- Imperialism: To what extent can foreign powers lay claim over sovereign states?

- Scientific morés: Who defines the moral codes for emerging sciences and technologies?
- Environment: What are the rights and responsibilities of nations that manage the earth's resources?

Indispensable as they are in everyday life, none of these issues has the well defined and undisputed foundations that are the hallmark of modern science. They are prone to controversy. Though judges can be called upon to resolve differences and their verdicts can be enforced, judges are often unable to convince conflicting parties that their verdicts are correct and final. How can we resolve these dilemmas or at least ameliorate the situation? Let us return to the tradition of exact science for precedence:

- No concept, no matter how widely popular, is sacred.
- No person, no matter how wise and knowledgeable, is a saint.
- No one presents one's beliefs as evidence of one's righteousness.

Strict observation of such seemingly simple rules in nonscientific cases is not easy. A conscious effort to adopt such an approach, however, should be rewarding and should help one better understand others.

It is, of course, naive to maintain that disputes between individuals, societies, countries, or economic blocs are the result of a lack of understanding. On the contrary, it is often conflict over material resources and thirst for domination that causes large-scale calamities. Nonetheless, the scientific world of the twenty-first century has managed to create a legal and international infrastructure to condemn, if not prevent, the primitive, brute logic which asserts that the strong can take the possessions of the weak. Such infrastructure impedes acts of aggression, or at least is expected to do so. Nowadays, an aggressor does not need to operate openly, but rather can remain hidden in a maze of international conventions and protocols.

Finally, I am aware that the aforementioned technique of reduction to basics has oversimplified the problem. After all, long before the formulation of exact sciences, man's inexact creations, such as sports, arts, music, poetry, literature, and commerce brought people together. Here, I only wish to point out that today's science, by all standards, is the most vigorous force behind the development of all societies. It is used by everyone. Logically, its value-free methodology could serve as a common language for dialogue amongst people. It is worth the effort, even though hermeneutic philosophers tell us that the task is not an easy one. No matter how hard one may try, one's intellectual horizon clouds the circumstances. In the words of Wilhelm Dilthey, the existence of other people has always been a scandal for objective thinking.

ACKNOWLEDGMENTS

I wish to thank my physicist colleagues, M. R. Hydari Khajepour and B. Farnudi, for fruitful discussions in preparing this essay.

DISCUSSION

Etienne Guyon: Many people talk about the relativity of science—what we believe today may be proven wrong tomorrow and so on. Are there things that are right or wrong and things that are not relative? We cannot, of course, foresee what is going to happen, but can you at least give definite examples that differentiate between what is right and wrong?

Yousef Sobouti: Professor Guyon, I think we agree on one point: the laws that we have discovered and attribute to nature have their domain of validity. We should not extend and extrapolate them beyond their domain of validity. Rather than say, "We knew something was right yesterday, and today we have proven it wrong," I prefer to say that what I knew was right yesterday is also right today. However today, with advanced technology, we may have finer observations that fall outside the domain of validity of the laws we knew. To explain these finer details, we may have to revise yesterday's laws and extend their domain of validity to accommodate the new observations. This is the way I look at nature, not in a black and white way to say that Newtonian mechanics were right up until yesterday, and relativistic dynamics threw it away. That's not true. Newtonian mechanics were quite correct within their limitations and were sufficient from the seventeenth to early twentieth centuries. But today we have further empirical evidence that finds Newtonian dynamics inadequate for their explanation. So we have revised it by the introduction of relativistic dynamics. In my opinion all findings of man throughout history are correct to a certain extent. And that certain extent has been changed through the years, maybe day-by-day, and still is being continued.

William Wulf: There are questions that affect science and are not reducible to scientific inquiry. What constitutes the legitimate application of science in a societal context is one question that is most often mentioned. But the method of deriving a scientific fact is not necessarily open to scientific inquiry by itself. The debate in many countries today about the use of stem cells and cloning is a current example, or a more horrific one is the kinds of medical experiments done by the Nazis in World War II. Would you elaborate your view on that?

Sobouti: My answer to your question on the use of stem cells is as follows: biology, biotechnology, and genetic engineering are still not exact sciences. They are not axiomatized yet. Once they become axiomatized, that is, once they are understood better than what we understand today, then many of the objections we now raise will simply be swept away.

7

Loving and Sharing Science: Pierre-Gilles de Gennes

ETIENNE GUYON
École Normale Supérieure, Paris

This presentation about Pierre-Gilles de Gennes, an intimate friend and a great scientist, recently deceased, should not be too surprising in a meeting devoted to opening the Gate to Understanding, as I aim to present some of the keys he used to open this Gate. This may be of special interest, as we are dealing with a scientific genius whom the Nobel committee which awarded him the prize in 1991, described as "a Newton of the twentieth century." At the same time, his achievements are accessible to others, as he liked to share his science as much as his love for science with others.

I will not present elements of de Gennes' long scientific career. This has been amply done in the months following his death. Rather, I will try to share some contents of his "tool box"—his set of keys—in order that others may have access to them. Some are quite general and are (or should be) taught in school. They establish the elements of rational judgment: order of magnitude, reference data, accuracy, and estimation of errors. These elements are often expressed as numbers or as quantitative ratios.[1]

[1] It should not be a surprise that in the midst of the French Revolution in 1790, priority was given by the elected deputies to establishing a standard of meas-

I will spend some time discussing the specific scientific tools that de Gennes used for his studies of magnetism, superconductivity, liquid crystals, polymers, soft condensed matter, and biology. Some recurrent items can be found.

- His interest in *order/disorder* effects in condensed matter as well as the essential role of *defects* in such structures. These defects should be avoided in some instances. In other cases, such as semiconductor electronics or properties of metal and alloys, they play an essential role.
- The use of *analogy* is a key that opens corridors between different rooms of science. Its use requires rigor and precise comparisons between different problems and should not be confused with loose metaphors. A mastery of analogy helped de Gennes build strong bridges between magnetism, superconductivity, liquid crystals, and polymers.
- *Interfaces* are another theme in de Gennes' research. Borders separate different entities, but they should also allow exchanges across them. It is possible in physics to transfer properties from one layer into an adjacent layer by such proximity effects. In such cases, original behaviors will emerge from this interplay. This type of geometry has been often considered by de Gennes in his work. It is also the basic principle behind the Nobel prize awarded to a "*normalien*" (as de Gennes was), Albert Fert, in 2007: by putting together stacks of very fine layers of a good conductor (copper) and of poorly-conducting magnetic iron layers, he was able to obtain some very anomalous magneto-conducting effects that are ubiquitous in today's memory chips of our computers and portable phones.

Quite metaphorically, we could say that these various properties—dialogue between order and disorder, the presence and role of defects, and the interactions at interfaces—are all elements of society! These scientific tools, having a large scope of application

---

ure—the meter or the kilogram—in order to provide a common reference that could be shared among all citizens.

and making recourse to mathematical treatments, should be analyzed in more detail in order to identify where and how genius enters into the work of de Gennes. Like a magic wand, he was able to transform a simple observation of phenomenon into a new field of research that others could build upon and that would lead to practical applications for many years to come.

Beyond these scientific aspects, I now want to try to identify some of the personal and social influences in his professional life. A first characteristic is the broader culture of which he was a part, where science is only an element of reference next to literature, art, philosophy, or history. The *École Normale Supérieure*, where he was educated as a student, was created to train some of the best students and to encourage cultural diversity, with an equal number of students in the colleges of humanities and science. In the sciences, we enjoy the benefit of a broad perspective and interdisciplinary approaches as keys for opening the gate toward new understandings. De Gennes often criticized the fact that our school programs and classroom curricula were too narrowly focused and left little room for such multidisciplinary opportunities.

A paradoxical attitude that great creators often have is to admit a certain bit of irrationality in their process of discovery that feeds imagination and, later, organizes creation. Why refuse intuition or serendipity if it results in discovery? I like to imagine de Gennes as a seventeenth century explorer of the New World, functioning as an astronomer, anthropologist, botanist, historian, and geographer. For de Gennes, an explorer's attitude is one of extreme curiosity to explore unknown territories of science as a leader, with an open mind and rejection of taboos and preconceived ideas.

This often teaches us a new lesson: the need to recognize and make use of one's errors. Teachers often do not dare say that they do not know something, or even worse, that they have made a mistake. This is clearly an essential part of culture and learning: to profit conscientiously from one's ignorance when in contact with others. Sharing is not just giving but also receiving, and de Gennes always listened carefully, regardless of the difference in age or education between him and his collaborators or students. Not only did he have a lot of charm and charisma when he expressed in the simplest terms what he had to say, he also liked to circulate in the ranks of the classes and listen very carefully to the questions and comments of the pupils. This attitude has to be promoted to open the Gate. We are not just there to open the "good book" of knowledge for others but to listen to what less scientifically educated individuals have to say.

An important social aspect he created in his laboratories was the team spirit he instilled. Rather than publishing with separate names, he would promote joint articles with no specific names to identify the authors: "Orsay group in superconductivity," "Orsay group on liquid crystals," etc. Such articles mixed contributions from experimentalists and theoreticians, young and senior members—even if their partial contributions were of different importance—with the largest part often coming from him. It was very stimulating and challenging for the younger members of the group, and I have also successfully tried to do it for my research groups.

I hope that these personal memories and reflections inspired by my friend Pierre-Gilles de Gennes will offer some keys to promoting science and leading to a better understanding among each other, with rigor as well as tolerance.

8

Successes in Building International Bridges through Science

NORMAN NEUREITER
American Association for the Advancement of Science

Today I want to present several examples of how science has helped build new bridges of understanding and cooperation between the United States and other countries for many decades.

My first example involves Japan and goes back to 1961. Japan was still rebuilding from the devastation of World War II, and the Cold War between the West and the Soviet Union was intensifying. At that time, a Harvard professor and famous Japan specialist, Edwin Reischauer, referred to the "broken dialogue" in U.S.-Japan relations. He was concerned over the breakdown in communication and lack of understanding between the intellectual communities in the United States and Japan. Japanese universities seemed increasingly sympathetic to the idealistic appeals of the Communists rather than the path on which Japan was rebuilding. Reischauer was appointed as Ambassador to Japan, the first Japanese language speaker to have that position.

Ambassador Reischauer wanted to fix the broken dialogue. In 1961, at his urging, Japanese Prime Minister Hayato Ikeda and U.S. President John Kennedy announced the creation of three U.S-Japan joint committees: a cabinet-level committee was to discuss economic issues, a cultural committee was to bring university scholars together, and the Japan-U.S. Joint Committee on Scientific Cooperation was established, the first of its kind in American history. The implementing agency for this committee in the United States was the National Science Foundation (NSF), which established an office in Japan to facilitate communication; and the Japanese responded by naming appropriate agencies to manage the program on their side.

The program moved very slowly at first; funds had to be appropriated in both countries, members of the joint committee appointed, implementing offices established, and acceptable scientific projects identified for cooperation. In early 1963, the program was just starting when I joined the NSF in Washington, and I became the first permanent U.S. director of the program.

There were a number of problems. It was not easy to find projects that could be truly cooperative because the level of science in Japan in terms of laboratories and equipment was well below the level in the United States. Secondly and unexpectedly, President Kennedy's science advisor, Jerome Wiesner, was concerned that science funded to achieve a "political" purpose was not subject to the same rigorous peer review process as other research projects, and hence might involve second-rate science. On the Japanese side, some professors initially seemed reluctant to become involved. Finally, too few American and Japanese scientists knew each other well enough to even think about cooperating.

To bring scientists in the same field together from the two countries, we funded many joint workshops in the belief that common interests and personal acquaintanceships would lead to joint projects. We financed a number of projects in common areas of competence, such as earthquake prediction, whale studies, and cancer epidemiology.

Ultimately, the program was recognized as a success in both the United States and Japan. Remarkably, it still exists today, although in a different form and without long-term funding for collaborative research. It has also served as a precedent and model for more specialized U.S.-Japan cooperative science activities that developed later, for instance, in medical sciences, in natural resources, and in deep-ocean drilling.

A second example in 1972 involved China. The Cold War was at its peak. The United States and the Soviet Union constantly threatened each other with large nuclear arsenals on both sides, although there was an ongoing U.S.-Soviet bilateral scientific dialogue that had encouraged restraint in both countries. Furthermore, the once close alliance between the Soviet Union and China had soured, changing the Asian geopolitical environment. U.S. contacts with China were essentially nonexistent. Except for some very sporadic and limited meetings of low-level officials that took place in Warsaw, Poland, there were effectively no contacts. In early 1972, President Nixon decided to try to change the U.S. relationship with China. The diplomatic history of his visit to China that eventually led to normalization of relations is well known. But what is not known is the role that science may have played.

At that time, I was working as the assistant for international affairs to President Nixon's science advisor. The U.S. government decided that in addition to the political change that we planned to raise with the Chinese, we would offer something of direct tangible benefit, specifically cooperation in science and technology. I was given the task of developing proposals that could be offered as part of a total diplomatic package.

Interestingly, the National Academy of Sciences (NAS) had established several years earlier a Committee on Scholarly Interchange with China. The committee had had little contact with Chinese scientists; nonetheless, it was a forum for people to consider what might be done if the political situation changed. Working with the staff of that committee, we produced about 40 initiatives for science cooperation in areas such as water and environment. These proposals became part of the package that eventually went to Beijing. Later, when diplomacy was finished and opportunities for cooperation began, the Academy, as a non-governmental body, was asked to take the first modest steps toward cooperation.

After the United States and China established diplomatic relations in 1979, cooperation began in earnest. Representatives of

19 U.S. technical agencies established the agreements that eventually led to America's largest international cooperative science program. Since 1987, up to 50,000 Chinese students per year have traveled to the United States, with about two-thirds of them interested in science and technology. At the beginning, almost 90 percent of them did not return to China and stayed in the United States to become university professors, work for U.S. companies, or start on their own as entrepreneurs. Today, Chinese students who finish their degrees as well as some scientists who have been in the United States for many years are increasingly returning home, as research conditions and financial rewards improve in China.

A new term has been coined to replace the concept of brain drain: *brain circulation*. This is now happening with Indian scientists who have studied and remained in the United States, following a pattern that began with Taiwanese and Korean scientists. This development has raised new policy questions as to the desirability of these foreign student and postdoctoral programs. As foreign specialists trained in the United States return to their countries and conduct research, the result is increased international competition in various fields of high technology. Concerns have been raised about America's long-term global competitiveness under these circumstances.

The answer is to attract more young Americans to study science and technology, not to keep out international students and researchers. Having close scientific colleagues in other countries is an advantage. It is the perfect basis for cooperation. Scientists with common educational backgrounds and personal knowledge of each other can readily cooperate in seeking solutions to many of the problems that confront the globalized economy—ensuring drinking water for growing populations, limiting emissions of greenhouse gases, developing alternative energy sources, and controlling emerging infectious diseases, for example. Unfortunately, our new U.S. visa policies and procedures since 9/11 are creating difficulties for international students and scientists to travel to the United States

Turning to a third example of international science cooperation, in 1967 I was the science attaché at the U.S. Embassy in Warsaw. My role was to contact members of the Polish scientific community and explore ways to cooperate, despite all the political problems between the countries.

We had an unusual advantage in Poland. After 1956, there had been a political change that permitted Poland to have closer ties with the United States. This included the purchase of large quantities of U.S. grain with Polish currency, which at that time was not convertible to dollars, but was available for joint U.S.-Polish activities in Poland. As a result, Polish zlotys could be used to fund cooperative scientific projects between Polish and American researchers. Many projects were developed, particularly in the agricultural and health sciences. Eventually, new projects expanded into other fields such as coal research and environmental protection. The volume of these projects eventually reached such a high level that the zlotys were exhausted in about a decade. However, since both governments recognized the value of these cooperative programs, they established the Maria Sklodowska Curie Fund. The Polish government contributed zlotys and the U.S. government provided dollars. That program continued for many more years. Currently, a generic science cooperation framework agreement exists between the countries. Unfortunately, the present agreement has no dedicated funding, meaning that each cooperating institution must find its own funds for participation in any cooperative project, which limits the extent of cooperation.

A fourth example of how science cooperation can build international bridges is much more recent and involves India. U.S.-India relations had been up and down for many years. An early idea was to have a formal agreement of scientific and technical cooperation, but it was recognized that such an agreement was not possible. Instead, a private organization, the Indo-U.S. Science and Technology Forum, was established to carry out scientific and technical cooperation. The two governments provided a modest endowment with rupees owned by the U.S. government. The endowment generates interest and the Indian government annually matches that interest, giving the organization about $1.5 million-2.0 million per year to catalyze cooperation.

In late 2000, I was appointed the cochair of the Forum. We have a secretariat in New Delhi that manages the funds and a small center in the United States to conduct peer review of proposals coming from both the Indian and U.S. scientific communities. There are about 50 events each year, including a large number of workshops across all fields of science, travel grants, and professor and researcher exchanges. We try to fund those proposals that show the greatest promise for creating real working collaboration. Since 2005, U.S. science relationships with India have deepened, a formal bilateral science agreement has been signed, and other proposals for cooperation have been considered.

Let me conclude with one final example, which could provide an opportunity for Iran today: the International Institute for Applied Systems Analysis (IIASA). It is housed in a former Hapsburg Palace in Laxenburg, Austria, a suburb of Vienna. The principal initial partners were the United States and the Soviet Union. Some 18 countries also joined and the research was focused on applying systems analysis techniques to big global issues. The early analysis of energy problems was considered to be very good, the research on population problems was of international interest, and the work on global climate change contributed to present global thinking on that issue.

Support for IIASA has waxed and waned over the years, and with the end of the Cold War, its nature changed. At present, it is no longer focused on building bridges between East and West but has taken on a North-South orientation, with more emphasis on developing countries as it continues to tackle major global issues. China, India, Pakistan, and South Africa have joined or are in the process of joining. A top IIASA priority now is a Global Energy Assessment, an examination of the global energy situation from many perspectives over several years. Iran was approached through one of its diplomatic missions about joining this energy project. To date, the Iranians have not responded. But participation in the energy project would be possible even if Iran chose not to become a full member of the organization. IIASA is another clear demonstration of how science can be a successful builder of international bridges between countries. Science is indeed a gateway to understanding. International cooperation in science and technology, which brings researchers together in the search for solutions to global problems in the natural world, can also be a powerful catalyst for improving relations in the social and political worlds. In any case, it is well worth supporting.

9

The Universality of Science: Examples from History

HOSSEIN MASOUMI HAMEDANI
Sharif University of Technology

The question of science's ability to promote mutual understanding among different civilizations, nations, and social groups is closely related to the question of the universality of science. Those who view science merely as an anthropological phenomenon, a product of western culture, a subculture shared by those who belong to a specific community, or a language game played by those who know the rules of this game and thus are called scientists have little to do with mutual understanding through science. From that point of view, science is one of the many factors that divide humankind. Of course, we can still gain an understanding through the study of those different scientific traditions or activities that are labeled "science" in their respective societies, but the understanding gained in this way cannot help us achieve common ground. The best it can do is to prepare us to accept the existence of others or to tolerate their existence in a world with different cultures—especially one with different scientific cultures. So my paper explores a single question: How universal is science?

The question of the universality of science can be investigated through three aspects: the epistemological aspect, the social aspect, and the historical aspect. These three aspects are somewhat related. Before discussing the main subject of my investigation, which is the historical aspect, let me say a few words about the other two.

The first aspect of the universality of science, the epistemological aspect, concerns the logical status of scientific theories and results: To what degree are scientific results and scientific theories acceptable to every human being? The answer seems to be straightforward enough. The universality of science is said to be due to the universality of its methods and the way its results are shared. The natural sciences are considered universal because they follow the scientific method and their results can be shared and understood by different human beings. Many schools of science philosophy define these two properties as "intersubjectivity" and "testability of science." At the root of this concept is the idea that scientific results can be communicated in an unambiguous way and tested by anybody who wishes to do so. As far as simple, down-to-earth facts of science are concerned, the explanation of scientific methods and results seems to be relatively simple. However, when it comes to more sophisticated theories and results, the task becomes very prohibitive. Scientific theories and facts are often expressed in the kind of mathematical language or theoretical jargon that is far removed from the vernacular and outside the experience of ordinary sensual perception. To understand a scientific theory and thus to be able to test it, one must belong to a defined scientific community, master the language used by its members, and share the methodologies specific to that community.

A way out of this dilemma is to state that scientific results are universally testable in principle. However, this statement does not solve the problem. One can always ask how this principle can be realized, and the answer brings us back to our starting point.

The German physicist and philosopher Carl Friedrich von Weizsäcker suggests another criterion that seems more practicable. According to him, the universality of science is measured in terms of the trust we place in science. In everyday life, the layman has little direct contact with pure science. All he knows of science is its manifestation as technological product, and the layman usually trusts the workings of those products. He pushes an electrical button; if the light does not turn on, he does not blame the science of electricity, but rather he thinks that something has gone wrong with the wires or switches. In this way he expresses his confidence in technology and, indirectly, in the science that lies behind this technology.

If we accept this criterion, we can say that science is universal because our trust in technology is universal. But the fact remains that most people know almost nothing about the science upon which their technology is based, so this kind of trust in technology can be equivocated with a magical view about the manner in which these products work. We can suppose that there is a spirit at work behind every instrument. That is why Weizsäcker compares this kind of trust with that which people place in religion and deities.

Therefore our trust in science-based technology is not equivalent to the idea of the universality of science unless it is combined with knowledge about how technological products function.

The theories of inter-subjectivity and the testability of science do not prove the universality of science but rather point to its *potential*. Science disciplines are potentially intersubjective, and scientific methods are potentially testable. These two potentialities can be actualized only through conscious human effort, and their realization depends on various factors that are not dictated only by the logic of science itself. For example, we can decide in principle between any two rival theories by doing a suitable experiment, which is usually called a crucial experiment. But in practice, to be able to perform an experiment demands many prerequisites. One has to have access to relevant technologies—to be able to build the necessary instruments—and these technologies are not always accessible. Historical science presents us with many examples of theories that have had to remain undecided because the relevant experimental technology or instrumentation is not available. The question of the speed of light—whether light travels at a finite speed or instantaneously—is only one example.

Thus, in its epistemological aspect, the universality of science expresses not a reality but rather an ideal. The potential for this ideal to become a reality depends on how much scientific knowledge human beings can possess. To put it another way, science is universal to the extent we want it to be universal. This brings us to the second aspect: the social aspect.

Even if we agree with Descartes that reason is distributed among human beings in the most equitable way, we have to agree with the French anthropologist Claude Lévi-Strauss, who stated that nothing is distributed more unjustly than science. The production and distribution of science is far from equitable, and those who find themselves at the consumer end of scientific activity do not benefit from it in a just way. Moreover, the division between those who understand science and those who do not understand science separates not only continents and countries, but also societies, even the most advanced societies. Paradoxically enough, the very same conditions that have brought the fruits of scientific activity to the layman have greatly contributed to this unjust situation. By this, I mean the link between science and technology and the emergence of techno-science. The fact that the creation of knowledge often results in the creation of wealth has put certain limitations on the availability of scientific information, and this development has been harmful to the cultural aspect of science.

I do not believe that everyone has to know everything. I do not believe that science is an ideology that has come to replace all other ideologies. I do not believe that there is a scientific solution for every problem or that science can put an end to every conceivable disagreement between human beings. Nevertheless, I believe that a certain amount of scientific knowledge and a certain perspective inspired by science are necessary for every citizen of the modern world.

This kind of scientific approach includes a sensibility toward concrete things, the ability to formulate relevant questions, and the search for the simplest ways to answer them. As far as my own country is concerned, I think that our educational system does not cultivate these qualities in our youth. Instead, it helps create a kind of scientific illiteracy, despite the large amount of science and mathematics taught in our high schools.

Fighting against scientific illiteracy is part of the social responsibility of every scientist. It is only by speaking with ordinary people and by explaining to them some basic scientific facts that science can recover at least a part of its lost universality.

Now I am going to speak about the third aspect of the universality of science, the historical aspect, and how the history of science can help us to share a more universal idea of science.

Modern science emerged from the scientific revolution of the seventeenth and eighteenth centuries as western Europe experienced a departure from some age-old habits of thought. To establish itself as a legitimate means, and even as the solely legitimate means, of investigating the natural world, modern science had to fight against many existing theories, concepts, and outlooks. This has given rise to a certain kind of historiography of science with the following characteristics.

- It represents the advent of modern science as a schism from all pre-existing modes of thought. Moreover this transformation is thought to have affected all disciplines in the same way.
- Anything that had existed before the scientific revolution is considered to be nonscientific, or at best as part of the prehistory of modern science, which began in the seventeenth century. For prehistorical science, historians often refer to Ancient Greece, but even the Greek legacy is divided into two parts: some theories such as the atomistic theories are seen as "precursors" of modern science, while other theories are considered to be obstacles to an accurate understanding of the physical world.
- Even those scientific theories that were created after the scientific revolution but were not compatible with mainstream science are considered to be vestiges of the past. For example, in

his attack against the Cartesians, Voltaire accused them of believing in occult practices and resorting to a kind of Aristotelism.

- In its more recent versions, and under the influence of the formalist or logical concepts of mathematics, this kind of historiography recognizes a separation between mathematics and natural sciences. The development of natural sciences and the development of mathematics are told as two unrelated histories. The mathematical aspect of modern natural science is shadowed by its experimental aspect, and the latter is reduced to a simple use of senses. This last characteristic is considered to be the main dividing line between modern science and the wild speculations of the ancients.

Thanks to the efforts of several generations of historians of science, this is no longer the history of science recognized by expert historians. But unfortunately, it still remains the story of science told in popular expositions and shared by many practicing scientists. The result is that science is represented in an ethnocentric and anthropological way. The only significant difference is that science is not a set of bizarre practices common to the members of an isolated and primitive tribe, but a rational universal practice that is nevertheless restricted to a big, powerful tribe.

In this kind of historiography of science, the birth of the new science is explained only by invoking some external elements. One example is the economic development and the emergence of new philosophies that occurred in Europe toward the end of the Middle Ages. In the first case, the history of science became a part of social and economic history; in the second case, a part of the history of philosophy.

This trend in the historiography of science has given rise to a reaction among non-western countries, with each country trying to rewrite its own history of science. Consequently, this field has become a battleground for nationalistic and even chauvinistic ideas.

Certain results of the new trend in the historiography of science help us have a more balanced idea of the universality of science.

- Science is not western in origin; notwithstanding the great role played by ancient Greece's contribution to science, there was no such a thing as a Greek miracle. In those disciplines at which the Greeks excelled—particularly mathematics and astronomy—they were deeply indebted to the scientific traditions of other cultures.
- The Dark Ages were not so dark. Without the scientific developments that took place toward the end of the Middle Ages, modern science would have been inconceivable.
- The role of Islamic civilization was not limited to the preservation of Greek scientific heritage and the subsequent passage of that heritage to Europe. Certain disciplines that emerged during the Islamic period and that have no counterpart in classical antiquity played a very decisive role in the constitution of modern science. Foremost among these was the science of algebra, which provided a more general concept of calculus and helped overcome the ancient dichotomy of continuous and discrete magnitudes.
- The scientific revolution did not affect all the branches of science in the same way. Some scientific disciplines, optics for example, experienced continuity throughout Antiquity, the Islamic period, the late medieval period, and even well into the seventeenth century. At least until Johannes Kepler, the history of optics followed the same principles that had been laid out by Alhazeni in the tenth century.

In conclusion, the universality of science cannot be guaranteed by its epistemological status alone. A truly universal science depends also on its accessibility to every man and woman and upon our recognition of its international origins.[1]

---

[1] Editor's note: Western science is based on the premise that the laws of nature are *universal* in that they are the same throughout the *universe*. To wit: a

DISCUSSION

Yousef Sobouti: Dr. Masoumi, you used the phrase "scientific illiteracy." Can you explain in simpler words exactly what you mean by that?

Masoumi Hamedani: In Iran, science is taught as if it has nothing to do with everyday life. It does not inspire curiosity in our children about the natural phenomena that surround them. Our children can solve any kind of conceivable mathematical equation, but they cannot give an estimate of the size of a phenomenon they encounter in nature. This is what I meant by scientific illiteracy.

Another facet of scientific illiteracy that is more visible and more or less global is the rejection of scientific theory in favor of cultural and religious dogma. An example might be a molecular biologist who believes in creationism—there are such people. It is not certain that science is compatible with all ways of life and beliefs.

Glenn Schweitzer: Would you comment on the impact of television on scientific literacy? Do you think that television has brought some great changes in the way scientists proceed?

Hamedani: In my view, Iranian television is good and possibly the sole medium that exposes our youth to certain aspects of nature, wildlife, and natural phenomena. But I have no ready answer to your question, and at the international level it requires a vast investigation.

---

stone falls identically in Tehran, Washington, or Tokyo. For the layman, and in particular for students in school, this may create a sense of universality in our world in spite of all conflicts and divergences between people. This is one reason why teaching science to schoolchildren is so important.

10

The Role of International Scientific Organizations

MICHAEL CLEGG
U.S. National Academy of Sciences

The science community has always been international in scope and practice. The basic propositions of the natural sciences are independent of culture and geography. For example, the theorems of mathematics or the principles of physics or the details of molecular biology are the same all over the world. Science communities have long worked to develop specialized languages to describe important concepts which provide a natural gateway to understanding between cultures and amongst nations.

What is it about science that is so special? Science is the world's most successful means of knowledge creation—that is, of understanding the relationships that govern the material world. Science deals exclusively with arguments based on empirical evidence or on theories about the relationships between material objects. Scientific propositions must be testable and subject to rejection by empirical observations. Moreover, the results of science must be subject to independent verification by others. Finally, the well-tested and enduring theories of science allow us to make limited predictions about the future. These properties are unique to science.

FUTURE CHANLLENGES

The next several generations will need all the resources of science to choose optimal pathways into the future. Humanity faces a number of unprecedented challenges that are in part driven by the rapid expansion of the human population and associated resource demands. The global human population was less than three billion in 1950 and expanded to six billion by 2000. It is projected to reach approximately 9.3 billion by the middle of the twenty-first century.

This means that in 100 years, human-driven resource demands will have at least tripled. The amount of arable land to support each person will have been reduced threefold, and many essential resources will be severely strained. Moreover, the by-products of human activity—such as CO_2 emissions—are changing the global climate and threaten to have uneven and possibly devastating impacts in some regions of the world. A short list of major twenty-first century challenges includes the following:

- climate change,
- water supply and quality,
- global energy transition,
- food production,
- emerging diseases,
- land degradation,
- ecosystem and species preservation, and
- equity and quality of life.

These challenges are global and do not respect national boundaries. In almost all cases, science can suggest potential mitigations or even solutions to these challenges, but to do so science must be able to present options based on the best current knowledge to decision-making communities around the globe. A fundamental goal of the international scientific organizations that are built on national science academies is to create a bridge between science and decision-making communities in every country (Clegg and Boright, 2007).

INTERNATIONAL SCIENTIFIC ORGANIZATIONS

The International Council for Science (ICSU): It traces its origins to the end of the 19th century but achieved its contemporary organization in the 1930s. ICSU is composed of a matrix of national members (about two-thirds of which are national science academies) and disciplinary unions. The stated mission of ICSU is to strengthen international science for the benefit of society. ICSU achieves this by (1) agreeing on the international language of science (nomenclature) and fostering international scientific cooperation in areas such as the maintenance of global databases and other research resources; (2) articulating global science projects such as the current International Polar Year; and (3) representing the science community with major U.N. bodies such as UNESCO.

The InterAcademy Panel (IAP): In the early 1990s, a global network of science academies known as the InterAcademy Panel on International Issues (IAP), was created. The precipitating event was a meeting in Cairo, Egypt, on the problem of population growth, where a white paper was issued by a number of academies. Today the IAP includes member academies from more than 90 countries around the globe. The goal of IAP is to build the capacity of member academies to advise governments on major science policy and these are uniform across the world. It is also a common practice for individual scientists to study in other countries and to attend international conferences aimed at widely disseminating new findings. Science progresses by debating the validity of new ideas and the empirical evidence that tests existing theories. This dialogue is not restricted by national boundaries. A global dialogue best achieves the advance of science.

The notion that science is universal is not new. Galileo, for example, devoted much effort to communicating with his contemporaries elsewhere in Europe as he sought to refine his theories and observations on falling bodies and on astronomy. Today, with rapid global communications, discoveries are communicated instantaneously and debated in scientific communities in all corners of the globe. IAP began by issuing joint statements on scientific aspects of global issues. The current list of IAP statements[1] include the following:

- population growth (1994),
- urban development (1996),
- sustainability (2000),
- human reproductive cloning (2003),
- science education (2003),
- health of mothers and children (2003),
- scientific capacity building (2003),
- science and the media (2003),
- access to scientific information (2003),
- biosecurity (2005), and
- teaching of evolution (2006).

IAP also has several programs and initiatives that address major issues and are aimed at assisting member academies by providing useful materials for decision makers in various regions of the world. The programs and initiatives for the period 2004-2007 are as follows:

- capacity building for academies,
- science education,
- health education for women,
- water initiatives,
- biosecurity,
- genetically modified organisms,
- access to scientific information, and
- natural disasters.

[1]See *www.interacademies.net/*.

IAP is now in the process of selecting new programs following its General Assembly meeting in Alexandria, Egypt, in December 2006.

There are analogous medical academy[2] and engineering academy organizations.[3] All three organizations work together toward the shared goal of providing the best scientific engineering and health advice to governments around the world.

The InterAcademy Council: In 2000, the InterAcademy Council (IAC) was formed. To quote from the IAC website:[4] "The IAC produces reports on scientific, technological, and health issues related to the great global challenges of our time, providing knowledge and advice to national governments and international organizations." The IAC is modeled after the National Research Council of the U.S. National Academies in that it assembles expert panels from throughout the world to produce high quality, in-depth studies of major science policy issues. To date, the IAC has produced the following reports on the importance of science capacity-building in every country: *Inventing a Better Future*, *Realizing the Promise and Potential of African Agriculture*, and *Women in Science*, all of which argue that the full utilization of human resources is essential for future success. The report, *Lighting the Way: toward a Sustainable Energy Future*, released on October 22, 2007, is typical in that the 15-member expert panel is composed of scientists from 12 different countries including Austria, Brazil, Canada, China, Egypt, India, Iran, Japan, Kenya, Russia, Sweden, and the United States. This broad international representation demonstrates that it is possible for scientists to arrive at consensus recommendations on the major issues of our time. These recommendations are meant to assist those in decision-making communities in all countries in selecting the best policy options to deal with one of the major challenges of the twenty-first century.

---

[2]InterAcademy Medical Panel (IAMP); see *www.iamp-online.org/*.

[3]Council of Academies and Technological Sciences (CAETS); see *www.caets.org/*.

[4]See *www.interacademycouncil.net/*.

The Academy of Sciences for the Developing World (TWAS):[5] TWAS is a member academy that elects its fellows from the distinguished scientists of the developing world and its associate fellows from the developed world. TWAS is located in Trieste, Italy and serves as the secretariat for IAP. TWAS has a number of programs aimed at capacity building and works closely with IAP, IAMP, and IAC to bring science to the policy-making communities of the developing world. Particularly noteworthy are (a) TWAS-based organizations of science ministers from the G-7 countries and (b) the Commission on Science and Technology for Sustainable Development in the South which includes heads of state from 36 developing countries. TWAS has made substantial progress in linking science to policy in the developing world.

Why are academy-based international organizations important? Academies typically include the most distinguished scientists of a nation and are thus regarded as authoritative voices on matters of science and technology. Because academies include the scientific leadership of a nation, they often have access to high-level decision makers and can communicate their advice effectively. Finally, academies are usually freestanding entities that are largely independent of government bureaucracies. This independence enhances the credibility of academy advice in the eyes of the public.

THEPOWER OF NETWORKS

The international scientific organizations described above have made progress in moving science academies from purely honorific organizations to service organizations. Most member academies of IAP and counterpart organizations welcome a greater role in bringing science to the solution of global problems. But the question of how to implement a global system of advice is challenging. Even though the list of issues given above is global in scope, solutions will need to be implemented locally, nation by na-

[5]See *www.twas.org/*.

tion. This follows because of national sovereignty. There is no international entity that can deal effectively with local issues like water allocation or local policies for alternate energy resource development. Moreover, optimal solutions will vary locally, depending on resource availability and the local impacts of phenomena such as climate change or water demands. To be effective, international science organizations must be able to reach local decisionmakers in the countries of interest.

Member networks like IAP and ICSU provide a solution to this outreach problem. Both organizations reach into most countries through their member academies (and in the case of ICSU, through its disciplinary unions). The challenge is to enhance the effective engagement of local academies and local science communities in communicating potential solutions to governments. Both IAP and ICSU recognize that a key to effective engagement lies in the creation of outreach mechanisms at the regional level. ICSU has responded to this challenge by creating regional offices that can craft outreach efforts to regional needs. IAP has incorporated a series of regional academy networks. The next steps for IAP are to use the regional networks as dissemination vehicles for IAC reports. Thus, for example, the regional networks will host workshops for academies on the policy recommendations of the IAC energy report to help local academies organize specific policy recommendations targeted to regional needs. This effort is still in formative stages, but the network concept has the potential to use trusted local science institutions to communicate policy solutions to national decision makers.

CONCLUSIONS

International scientific organizations have an important role to play in assisting nations as they navigate the many challenges that will confront humanity over the next half century. The notion of public service is an old one in scientific communities but has assumed a new urgency as we enter the twenty-first century. A vision of how science communities can be actively engaged has emerged and is in the process of implementation. There is reason to hope that local academies can serve to assist governments in crafting appropriate solutions to resource and other challenges.

DISCUSSION

Norman Neureiter: I am concerned about international organizations in their dealings with science issues. They all seem to be starving for money and resources. Do you agree with that? Is your academy effort, which is really quite impressive, trying to focus its efforts in some way?

Michael Clegg: Unfortunately the landscape appears quite bleak, and I think the names of these organizations don't help, because the names convey very little effective information. To the outside observer it appears to be a confusing landscape. However, the reality is that there is effective differentiation. The IAP, CAETS, and IAMP work closely together. In fact, IAMP and IAP, which are both headquartered in Trieste, share a common agenda, and we hope to involve the engineering community more in this effort.

The ICSU programs manage languages used in science by managing its data bases and nomenclature. They provide a mechanism for the international science community to interact with the United Nations (UN) organizations, primarily UN Educational, Scientific, and Cultural Organization. Also, ICSU articulates major global research projects, such as the current global polar year. All of these activities are very important, particularly the language in commerce of science. If ICSU did not exist we would have to create an organization to make international communication in science effective and smooth. The IAP, IAMP, and CAETS efforts are aimed at empowering science academies to use the three organizations to deliver important messages at international levels.

They are all nongovernment organizations. They are not parts of national governments. The IAP receives substantial financial support from the Italian government, however. There are efforts to work effectively across the boards, so the ICSU leaderships sit on the governing board of IAP, and both sit on the governing board of IAC. Thus, there is cross communication, and the effort to use IAP as a delivery mechanism for IAC reports is an important step in the direction of unification of these different efforts.

Now let's turn to the African Science Development Initiative (SADI). This is a project which is operated by the U.S. National Academies. It has been funded for a 10 year period by the Gates Foundation, and the goal is to strengthen science academies in Sub-Saharan Africa. We have worked primarily with seven academies in Sub-Saharan Africa. Senegal is one of these academies, and next month in Dakar we will have an annual meeting to assess the project. This project has completed its third year of the ten-year period, and the initial results of the effort are just beginning to come in.

For example, the South African Academy of Sciences recently released a major report to government on the relationship of nutrition with HIV. This happens to be a very controversial subject in South Africa because the minster of health in South Africa had been advocating nutritional approaches to dealing with AIDS. South Africa has the highest HIV burden of any country in the world; approximately 12 percent of the population is HIV positive, which is a huge burden for the country to carry. The scientific community does not believe that nutritional treatment offers any solution to AIDS, and the South African Academy issued a report that basically arrives at those conclusions. This was aimed at influencing policymakers to approach the AIDS problems from a more informed perspective.

In other areas, the Nigerian Academy has been doing workshops on safe blood products. The Ugandan Academy is in the process of developing its first report. So this project is in its early phases, but it is aimed at a key objective related to the larger theme that I mentioned: using academies because they represent the scientific leadership of the country. But typically that leadership has political access by using academies as a way to strengthen and empower science communities.

Yousef Sobouti: Is there any assessment of what portion of the recommendations of interacademy circles have been adopted by the policy makers?

Clegg: That's an excellent question. I think that in the initial IAC reports, because of how they were released and propagated, there was a fairly high impact. The first IAC report was released at a workshop chaired by Kofi Annan at the United Nations. It is a message of the importance of science capacity building. Most academies are very happy to carry that back to their own governments because obviously it speaks to the best interests of the work of the science community as well. The food security in Africa report was a very important report that has not been disseminated as effectively as we would like it to be. It has not had as large an impact as it should have. There is a serious effort to disseminate the report on women in science through academies.

We hope to use this workshop mechanism at a regional level to engage academy leaders region by region, because the complexity of the issues varies from one region to the next. So we can tailor the approaches and recommendations to local needs and engage academy leaders and decision makers in a way that makes the recommendation penetrate governments. But I would say that the jury is still out on this question.

REFERENCES

Clegg, M. T. and J. P. Boright. 2007. Adapting to the future: the role of science academies in capacity building. International Journal of Technology Management. UK.

InterAcademy Council. 2007. Lighting the Way: Toward a Sustainable Energy Future. Amsterdam: InterAcademy Council. Available online at *www.interacademycouncil.net/?id=12161*.

InterAcademy Council. 2006. Women for Science. Amsterdam:

InterAcademy Council. Available online at *www.interacademycouncil.net/?id=11278.*

InterAcademy Council. 2005. Realizing the Promise and Potential of African Agriculture. Amsterdam: InterAcademy Council. Available online at *www.interacademycouncil.net/?id=8927.*

InterAcademy Council. 2003. Inventing a Better Future. Amsterdam: InterAcademy Council. Available online at *www.interacademycouncil.net/?id=9988.*

11

Science and Culture

BERNARD MAITTE
Lille University of Science and Technology

Every day we use tools that have been produced as the result of extensive scientific research. Production methods have changed considerably, thanks to new technologies in chemistry, electronics, computer-aided design, materials, biotechnology, and food science. As science launches into an ever more detailed exploration of the infinitely small, so it continues to investigate the infinitely large, providing us with the means to cure and eradicate disease depleting the planet's resources and irreversibly damaging the environment. The remarkable advances made in biology have not only resulted in healthier births and babies, but have also been instrumental in fostering experiments on embryos, clones, and genetically modified organisms.

SCIENCE, CULTURE, AND THE HONEST MAN OF OUR TIME

ientific applications have led to significant changes in our everyday lives, to national economies, to our social aspirations, and in our personal impressions of the world. They have also given rise to profound ethical questions. In an era when science has such an impact on every aspect of our lives, most people, including a large percentage of those involved in teaching science, find it difficult to keep up. Science remains an external, widely misunderstood field. It is the domain of specialists who live in a world of abstraction and rigor, and only those with highly specialized knowledge have the keys. In addition, it is difficult to understand the real purpose behind the flow of technology or which societal needs drive these innovations. We are currently in a period of rapid transition from a society focused on mankind's future to a society that values objects and means but has not thought through, expressed opinions about, or made decisions regarding the future of the human condition. The gap appears to be growing between science and culture, regardless of the following definitions we give to this polysemous term.

- Culture as the intellectual product of an era: scientific activity is highly intellectual, but it is also extremely compartmentalized within narrow disciplines. Researchers must remain current in their areas of expertise, but often are unaware of the historical theories and concepts behind their modern science; they delay exploring underlying meanings to a later date. Yet within its own boundaries, science makes an important, albeit limited, contribution to culture.
- Culture as the total sum of a society's technology: there is a daunting barrier—for reasons as much psychological as intellectual—between those who create technology through science and those who consume the technological products of science.

The paradox is as follows: even though scientific research leads to the technological products that elevate the standard of living for all members of society, the elements of science—symbolic language and specialization—are too esoteric for the majority of the population to understand. Unfortunately, the education that we provide to our youth is too often unable to (a) awaken a sense of curiosity for scientific and technical questions, (b) keep abreast with developments in science and rapid changes in its applications, (c) understand its implications, or (d) present science as a living thought process that takes the risk of questioning itself. Science does not, in fact, form any part of the culture of the honest man of the end of the twentieth century.

I would like to show that this is a recent development, that it has developed in the wake of historical evolutions, and that it is due to the way science is practiced, taught, and disseminated.

MODERN SCIENCE ITS FOUNDING FATHERS

Ancient and medieval science was indistinguishable from philosophy. It aimed to promote global understanding of the universe regardless of application and practical utility, and it attempted to articulate an intellectual impression of the natural world. Aristotle, who had a determining influence on the development of science, believed that knowledge of the natural world was based on the senses, even though perceptions can be misleading or fallacious; we understand how things happen better than why they happen.

Science cannot be constructed on sensory perception alone; it must be built on reason, which requires an "inductive-deductive" method. Our senses allow us to perceive facts that can be integrated into generalizations, even when they appear contradictory. Through the intellectual exercise of induction, we should observe phenomena, infer substance from them, and then deduce that the observed effects indeed arose from the substance inferred, thus providing logical explanation. The scientific world is based on human observation, and intellectual prowess allows the human mind to work out relevant principles.

Physics is the science of intrinsic quantities, and mathematics is the science of discrete and continuous qualities; both lead us to metaphysics. This is the science of existence that allows us to deduce principles and explanations without concern for application, and science originating from Islamic countries added a great deal to this concept. Ibn al-Haytham concluded that contradictory explanations may also be tested using Aristotle's formal rules of logic. Experience became one of the categories of proof, and the hypothetical, deductive experimental method was born and developed. The Latin science of the Middle Ages dropped this crucial innovation, and it lay forgotten until the Renaissance, an urban civilization marked by the development of trade, in which counting, measurement, and application were essential.

At the time, Galileo and several of his contemporaries believed that the world is written in mathematical language, and they began to develop theories of modern science that borrowed substantially from Islamic science. In a world too complex to be studied as a whole, Islamic science isolated facts and discarded conditions considered to be superfluous, thereby reducing the properties of reality. It expressed chosen facts mathematically, developed hypotheses, tested them against calculations, and presented new facts for observation. Experience allowed theoretical deductions and predictions to be validated.

Science delayed an overall understanding of the universe. Only provisional approaches relevant within the framework of reductionism were possible. It could say yes, but not if or how systems and objects developed. The founding fathers of modern science set out three objectives: to arrive at an understanding of the world that surrounds us; to become like the masters and owners of nature; and to secure progress, push back the boundaries of ignorance, fight against superstition, and contribute to happiness. These three goals were sought simultaneously by seventeenth-century scientists, who believed they had demonstrated that the world was a huge clock and who described the universe in mechanical terms.

During the Age of Enlightenment, the same framework was retained and expanded. It proved that mechanics were insufficient to explain life processes. The nineteenth-century intelligentsia looked for ways to unify science through energy and electromagnetism. The beginning of the twentieth century brought fundamental changes in the way we thought about the world. Throughout all this, science remained in perfect harmony with culture. Is it possible to separate the philosopher from the scientist when we think of Descartes, Spinoza, Leibniz, Diderot, Kant, or Einstein?

The adoption of the scientific method, the professionalization of research, and the growing determination to focus on applied science inevitably led to the separation between disciplines, which became increasingly narrow over time. New characterizations of the world can thus be built on local knowledge developed within a narrow discipline, and an ever greater number of applications can be deduced. It can also be a weakness, since disciplines are often compartmentalized to such an extent that sometimes researchers find themselves poorly informed about issues that lie outside their expertise. The main activity of these researchers, as we said earlier, is to keep current with advances in their specialties. They may postpone the search for underlying factors until later, unaware of all historical developments relating to their own discipline and where their most recent work fits in. Contemporary scientific practice therefore may be forgetful, with the projects developed by the founding fathers of modern science having reached their limits.

DEVALUATION OF TECHNICAL RATIONALITY

The preceding analysis does not concern technologies, which have been developed since prehistoric times. Early technology was the result of mankind's determination to create tools to feed and protect himself, to improve trade, to save or increase his strength, and to dominate. In other words, it responded to human aspirations.

Modern science and Islamic science drew the tools and processes necessary for experimentation from these technologies, and thereby fundamentally affected the way that science developed. The history of science from the seventeenth through the twentieth century shows that the development of science and technology was based on a constant international exchange of knowledge across disciplines. The steam engine was invented by technicians, while the determination to improve the output of these machines was based on thermodynamics. Anything "heavier than air" was declared unachievable by scientists until a few wizards—

sometimes at the cost of their lives—managed to take off for a few meters in their "strange flying machines." This paved the way for major research into aerodynamics and ultimately led to the development of aviation.

Such examples are numerous. Scientific rationality was always fed by technical imperialism that provided the principles of order and economy, with technological rationality arising from observation, trial, error, modification, and dexterity. All the while it remained highly dependent on the tool's place in society, which developed into ingenious combinations to address the deep-rooted needs in human society.

Today these low-tech cultures have been undermined, at least in western countries. The possibilities they offer are considered inferior to advanced scientific developments and, with the restructuring of certain economic activities, they have tended to disappear. Low-tech cultures have been replaced by "new technologies," a neologism which was mistakenly attributed to the concept of "leading edge, modern, and complex techniques." This was to emphasize that their relationship with science was different and could be developed directly from the outcome of scientific theories.

But isn't it dangerous to introduce new technologies without taking human experience into account? Shouldn't we try to make sense of their intrusion before disseminating them widely and attempting to adapt users to them on a wholesale basis? Isn't it time to stop thinking of them as the sole reference point for human development? Hasn't their introduction led to cultural changes much faster than the changes in collective memory, resulting in inappropriate and even dangerous behavior at the individual level? Hasn't our generation alone been through more changes in terms of production methods than the 10 preceding ones? How can we metabolize human experience in such conditions?

These questions are not among the issues being debated today. As yet, no country has managed to introduce a real debate with respect to the scientific and technical choices it makes.

NEW TECHNOLOGIES AND THE LEVEL OF CULTURAL PROGRAM IN SOCIETIES

Many thought that new technologies could be introduced without consideration for the economic, social, ethical, or political conditions necessary to assimilate them and benefit from them. It was thought that society's cultural progress could be kept separate from its technological advancement. Only now are we beginning to understand our mistake. People in France are often trapped in a contradictory and emotional duality whereby faith in progress is set against the fear of its consequences, the assimilation of industrialization, and the loss of cultural identity.

Is it possible to move away from these false debates, to speak out freely, and to extend our options, making a clear distinction between civil and scientific authorities? Isn't it time to break away from current practices and to value instead "the slow man?" These are the words of people who take the time to think about where they are going, who go against the aims inherent in the introduction of technologies, and who instead adapt them to projects that they themselves fashion for their happiness. Shouldn't we take the time to transform the technological object into a technical tool, which implies adapting and transferring our human experience? Both Bell and Adler, the two inventors of the telephone, were unaware of each other's work, with one publishing in the United States and the other in France; and yet each had imagined only one possible application for the invention, which was to listen to concerts at home over the telephone line. It was only with time and practical experience that other uses were developed. Examples of this kind of situation are numerous.

According to Condorcet, moral progress and scientific progress go hand in hand. This implies that the more technically developed a society, the more it needs technicians with a high level of cultural awareness. However, Condorcet's optimistic view hardly appears to be the case at present. The number of specialists has increased, but we no longer take the time to value knowledge other than scientific knowledge, and we are gradually losing individual and collective experience.

A less superficial approach to new technologies indicates that they manage to penetrate the social and industrial fabric when they correspond to the needs of companies, people, and their cultures and when they are well-adapted, thanks to the know-how of technicians. Developing a more sophisticated industrial tool from a basis of traditional techniques is in keeping with the experience of societies. The new tool makes sense to the society that created it; and from this, another, more coded and more symbolic man-machine relationship can be developed. These observations indicate that it is vital to safeguard technical cultures, to transfer human experience to new technologies, and to develop them in line with the relevant culture. If technical memory disappears along with those who hold its secret, and if it is not transformed into another form of culture, the thread of its sense will be broken, resulting in impoverishment and the need to begin again from scratch. This poses a number of problems for education.

GIVING NEW DEPTH TO SCIENCE TEACHING

The difficulties currently faced by scientific practice with respect to the aims assigned to it by its founding fathers have resulted, to some extent, in a certain loss of the rationale for science education. In its current form, science teaching has both unquestionable strengths and manifest weaknesses. Its strengths include training in the use of concrete knowledge, providing specialized knowledge, and grounding in disciplinary expertise. Its weaknesses, on the other hand, are that it is presented too dogmatically, is too compartmentalized within impermeable disciplines, and does not use the qualities of observation, manipulation, and experimentation enough.

I have a dream of a teaching approach that is rooted in disciplinary knowledge and at the same time placed in its historical context. I have a dream of a teaching approach that would include debate and controversy. I have a dream of a teaching approach that would be open to cross-discipline ideas. I have a dream of a teaching approach that would be formed via the exercise of scientific thinking, fervent thinking that observes, builds hypotheses, tests hypotheses, gives up, tries again, takes risks, invalidates, confirms its theories, and questions itself. In other words, we need thinking that provides us with access to partial representations of the world, as we can never exhaust reality, and opens up other ways to access nature, other kinds of thinking and action that are nonetheless relevant representations developed between the limits of validity that we are able to define. Learning to think through a method that continually conducts research and challenges the best established certitudes and the results considered as certitudes in order to gain satisfaction of understanding would be truly motivating. This would be a method that refuses all hegemonic learning models.

How can we attain this without recourse to a practical approach? Science teaching should aim to develop a new humanism, training the honest man of our time for whom science would no longer be an unknown. We should not lose sight of the fact that it is essential for those who are intent on scientific careers to base their teaching on disciplinary knowledge. All learning is built on concrete as opposed to didactic knowledge. We need to learn how to handle the tools, to understand what they represent, and to appreciate the capacities of each of them. For this, the role of the master is primordial.

OTHER WAYS TO PLACE SCIENCE IN A CULTURAL FRAMEWORK

Learning is not limited to teaching, however. We all have spontaneous impressions, initial impressions formed by society or scientific information conveyed by the media and its extensions to a nonspecialist public. We need to be aware of these parameters. Newspapers, radio, and television generally present a number of disparate and specific facts to the public that celebrate successes and performance and focus on spectacular results, or on the contrary, expose dangers. They call on experts who readily go beyond their specific fields of competence and voice opinions according to their intellectual and material interests and ideologies. Often they use authoritative language or, instead, sidestep the issues by taking refuge behind the principle of precaution. Once again there is little place for context, issues, debate, or perspective on the choices that people must be made to understand, the same people who, at the end of the day, decide on which applications should be developed and the goals and purposes of those applications.

The popularization of science addresses a public that is already motivated. At the same time, it leads to neither the practice of science nor its true understanding. As I said earlier, this requires long training, constant effort and the perseverance of the education system. Popularization utilizes communication techniques that attempt to illustrate and that use analogies, metaphors, and models, producing somewhat distorted visions as a result of the self-representations of the people being addressed. Perhaps paradoxically, this helps to increase the existing gap between science and the public.

On the other hand, for a discipline that is essentially critical in its methods, there is little external criticism of science. Admittedly, there are very few who are able to criticize it. But critical feedback should mainly assess the relevance of the applications and launch debates on their implications for society. Perhaps this criticism could actively contribute to putting science into the much-needed cultural framework.

Other learning strategies could be introduced. These are partially and hesitantly used by institutions that present science to the general public. Education would benefit from a policy that includes the following four interdependent levels:

- The development of the motivation for and the pleasure of science discovery by revitalizing the concept of personal experience through observation, touch, testing, failing, criticizing and symbolization,

- A holistic approach that addresses all aspects of a topic, including historical, scientific, technical, artistic, and literary data in order to contextualize science and to put it into context with other fields of culture,
- The discussion of issues and debate about scientific applications which affect society,
- The development of documentary projects and functional networks that broadcast and share these approaches throughout cultural institutions.

By making use of the terms *invent*, *create*, *form*, and *activate,* we can encourage the public to learn intelligent scientific thinking, ask questions rather than offer answers, and debate science and its applications. However, the effectiveness of contemporary science masks, sidelines, and undermines the work required to develop true understanding. Education based on disciplinary knowledge needs to teach how the results obtained by science can be used and contribute to taking further the question, "What does it mean?" It must place the question of overall sense and meaning squarely back in its central position. Education must once again make mankind and its aspirations the focal point of its concerns.

Only then can we discuss the key question, "Must everything that is technically possible be done?" To answer this question, it is important to make scientific thinking a major element in the process, thinking which allows us to develop a rational study of the world and to modify it, but also thinking that is nourished by dreams, imagination and utopia if it is, in its turn, to transform dreams, imagination and utopia. I am talking about thinking which, while not sufficient to guide human affairs, as some mistakenly believed, is essential to understanding them. For this reason, this form of thinking must become part of the culture of our time, in the same way that culture must become part of science.

12

A Half Century of Successes and Problems in U.S.-Iranian Cooperation in Science, Engineering, and Medicine

GLENN SCHWEITZER
The National Academies

During the last half century, colleagues in the United States and Iran have undertaken a number of important cooperative programs in science, engineering, and medicine. Footprints from these efforts are embedded in laboratories, educational centers, and other institutions in both countries. During the same period, however, hundreds of programs have been proposed and even started that did not succeed due to political difficulties, lack of financial support, and a host of other reasons.

As is well known, bilateral cooperation reached a high point in the 1970s—educational exchanges, joint research projects, technology-oriented activities of multinational companies, and other forms of interactions. Currently, bilateral cooperation is at a very low level, although the attention in Washington and Tehran devoted to this particular workshop is quite impressive. There seems to be widespread optimism that science can indeed become a gateway to understanding.

Thus, an important objective of this workshop is to reinvigorate interest in both countries in cooperative endeavors that can benefit international science while also improving the atmosphere for development of better bilateral political relations. Central to this effort is overcoming political barriers that prevent or discourage cooperation.

This paper describes a few examples in U.S.-Iranian cooperation in recent decades. The review of past efforts and the obstacles that have been encountered should provide useful background for related discussions about future activities. Such forward-looking discussions should take place at this workshop.

My presentation emphasizes events that have been supported by the governments and by well-established nongovernmental institutions in the two countries. Of course more limited efforts by individual specialists may often be the most effective form of cooperation. And the importance of multilateral activities increases during times of political turmoil. But the symbolic value of bilateral activities involving well-known institutions should not be underestimated.

This paper provides a Washington perspective on the character and impacts of bilateral cooperation. The programmatic examples add specificity to the presentation. During the discussion, I hope that participants in the workshop will add Iranian viewpoints on these and other types of cooperative activities. Many participants have firsthand experience, and their views are important.

We should be able to extract significant lessons learned from the activities that are singled out for attention. Some events have had positive effects in advancing science and in building bridges among specialists with common interests. Others underscore the importance of designing programs in ways that reduce the likelihood that they will run afoul of political impediments to cooperation, either before initiation or during implementation.

The review should help document the important role played by advocates of cooperation in both countries who are prepared to sustain their advocacy efforts over extended periods. In many cases, their efforts have clearly demonstrated that cooperation is feasible and important, even under difficult political conditions. Of course in highly sensitive security-related fields, cooperation has been and will probably continue to be off limits regardless of the mutual interests in cooperation of nongovernmental institutions in both countries.

Beginning in 1952, the United States launched the Point Four program of technical assistance in Iran. For more than a decade this program involved many Iranian institutions and brought to Iran more than 400 American specialists in a variety of fields. More than 4,800 Iranians worked for U.S. foreign aid organizations. Tehran University, the Technical College in Abadan, and other education and research centers in different parts of the country were active participants in the program. During this period, Pahlavi University, renamed Shiraz University many years ago, was to a considerable extent patterned after American higher education and was populated with visiting American professors. Indeed, today it is still referred to by some graduates as the American university (Amuzeger, 1966; Bill, 1988).

In the early 1960s, the United States sold to Iran a 5-megawatt nuclear research reactor located on the premises of Tehran University. Also, the U.S. government provided about 10 pounds of enriched uranium to fuel the reactor, which reached the state of criticality in 1967. Originally the reactor was operated by the university. Shortly after its installation, control was turned over to the newly established Iranian Atomic Energy Office. This activity was the focus of intergovernmental discussions on cooperation in developing 30,000 megawatts of nuclear power capacity in Iran, a target that apparently has persisted within the Iranian government until today. Related to the early interest in nuclear reactors, 150 Iranian nuclear engineers were trained at the Massachusetts Institute of Technology with other engineers trained in Western Europe. Presumably, the complex where the research reactor is located continues to provide training for Iranian students in a variety of fields with nuclear applications, including medicine, agriculture, and electronics as well as in scientific research. At present, enhancement of Iranian capabilities in nuclear science and engineering is a contentious international issue, as is well known.[1]

---

[1] See, for example, *www.nti.org/_research/profiles/Iran/3119_3268.html* and *www.workers.org/world/iran-nuclear-0324;* both accessed September 20, 2007.

During the 1960s and 1970s, the U.S. private sector became intensely interested in the industrial development of Iran. Economic globalization was in its early stages, and Iran was considered a lucrative market for American products and services as well as an excellent training ground to hone young talent in advanced technologies. The Iranian government entered into contracts with American and other international companies to help develop the petroleum sector and to enlarge the irrigation and hydro power infrastructures. This water dimension of agricultural and industrial development was at times compared to the development of the water resources in the southeastern United States under the auspices of the Tennessee Valley Authority (Bill, 1988).

In the oil sector, a western consortia of companies operated Iranian facilities, but Iran retained ownership. As would be expected, the financial arrangements associated with activities in this sector were of great interest. Indeed, the finances sparked controversy in Iran and internationally (ibid).

Also, prior to the Revolution, sales of American military equipment to Iran grew rapidly, with annual sales reaching billions of dollars in the early 1970s. As Iran developed capabilities to absorb the advanced technology associated with military equipment, local efforts probably had significant spin-off impacts in strengthening the civilian sector as well. Some benefits eventually may have turned into liabilities, however, as Iranian facilities have struggled to find embargoed spare parts for maintaining military equipment in working order, thereby diverting time and talent from focusing on enhancing civilian technologies (ibid).

Until the late 1970s, tens of thousands of Iranian students traveled to the United States to obtain degrees in many fields. Engineering was the most popular topic, with more than one-half of the Iranian students in the United States studying engineering. In the 1970s, the number of Iranian students at American universities and colleges reached a peak of 50,000, with most financed by their families or other private sources. The number of students then steadily declined to a level of 2,400 in 2006-2007.[2]

These education experiences seem to have had a profound influence on the development of universities in Iran. In some cases, student activities led to sustained linkages between American and Iranian universities. Iranian graduates often took home their experiences in the United States. Some tried to shape approaches at Iranian universities to conform to models they had seen in the United States. According to many reports, most students had positive impressions of American approaches.

Some Iranian students have remained in the United States. Today Iranian-born scientists and engineers are making important contributions to the research and technology efforts of American universities and companies. For example, Stanford University's Department of Electrical Engineering considers graduates of Sharif University of Technology as the best prepared electrical engineering students undertaking graduate studies at Stanford and points to a number of former students from Iran who now hold key positions within the American academic community. And when there are tremors of earthquakes in the San Francisco area, Iranian engineers from the University of California at Berkeley are often prominent among the advisers to state and local governments.[3]

A specific proposal that offered considerable promise for developing the scientific capability of Iran was set forth in 1975 by The Rockefeller University to assist Iran in establishing a modern biomedical research center. The team from The Rockefeller University, which has been the home institution for a number of Nobel laureates, concluded that an effective new research institute could be established in five to ten years. Unfortunately, this proposal was not implemented, as there were many claimants on limited financial resources, which shrank as oil prices tumbled. Then, within

---

[2] Information obtained from Institute for International Education, New York, New York, September 15, 2007.

[3] Information obtained from Department of Electrical Engineering, Stanford University, September 15, 2007.

several years the receptive political situation was rapidly transformed (The Rockefeller University, 1975).

Nevertheless, in recent years we have witnessed a significant strengthening of Iranian capabilities in the field of biomedical research—at Tehran Medical University, at Shaheed Beheshti Medical University, and at the Pasteur Institute, for example. These institutions maintain limited but nevertheless strong ties with American institutions despite political problems. This is indeed an area for cooperation that will benefit populations in many countries.

Moving forward to 1994, the U.S. Congress passed the Iran-Libya Sanctions Act (ILSA), which complicated the carrying out of bilateral scientific exchanges. Restrictions were placed on all American organizations that desired to carry out sustained activities with Iranian institutions in science and technology. The key test is whether the activity could be interpreted as involving a "service" provided either by the American or by the Iranian partner. Even publishing Iranian papers in American journals was initially interpreted as providing a service since the journals inevitably provided some editorial service in preparing papers for publication, as discussed below.[4]

Also in 1994, the Iranian Academic Association was established in New York City to provide a discussion forum for Iranian scientists from academia and industry through conferences and workshops on developments of broad international interest. The association has sponsored a series of workshops and meetings in Iran and in the United States on topics such as earthquake response, automobile accidents, the petrochemical industry, and environmental pollution. However, during the past several years, their activities seem to have slowed down.[5]

In 1999 and 2000, the leaderships of the Iranian Academy of Sciences and Academy of Medical Sciences and the leaderships of the U.S. National Academy of Sciences, National Academy of

---

[4] See, for example, Office of Foreign Assets Control, Treasury, 31 CFR Parts 515, 538, and 560, December 15, 2004.

[5] See *iaa20.tripod.com*; accessed September 20, 2007.

Engineering, and Institute of Medicine met in Washington and Tehran. These meetings led to the launching of a program of scientific workshops and individual exchanges that have taken place during the past seven years. The activities have been particularly important in demonstrating that cooperation is possible during a period of deteriorating political relationships. The programs have been carefully designed to avoid legal and political difficulties.

In 2001, several American organizations, including the three components of the U.S. National Academies mentioned above, were working with the U.S. government to exempt scientific organizations from the provisions of ILSA that restricted normal scientific exchanges. A general license was to be issued by the U.S. Department of the Treasury to this end. But just as the proposal for exemption was on its way to the White House for approval, terrorist attacks were carried out against the United States on September 11, 2001. The proposal was cast aside by the U.S. government as possibly threatening the security of the United States.

At about the same time, consideration was being given in Washington to establishing a variation of the Fulbright program that would benefit American researchers interested in living for an academic year in Iran, to be followed by Iranian scholars interested in temporary stays in the United States. The Fulbright program, which supports exchange visits of graduate students and researchers in many fields, has for several decades been one of the most successful international programs supported by the U.S. government. Even though there was considerable interest in the American academic community in participating in such a program, the events of September 11 immediately dominated the thinking of political leaders. The proposal was quickly put aside and has not been revived.

In 2003, the frustration of American scientific organizations with the restrictions on editing and publication of articles by Iranian scientists culminated in protests to the U.S. government. This effort was led by the Institute of Electrical and Electronics Engineers (IEEE). With strong backing from a number of U.S. professional societies, the IEEE succeeded in persuading the Department of the Treasury to issue a general license, which now permits publication of Iranian articles in American journals and preparation of joint reports without the previous requirement of seeking an individual license each time such a paper or report is being considered (Office of Public Affairs, 2004).

The Librarian of the U.S. Congress, James Billington, visited Iran in 2004. He was the highest ranking government official to travel to Tehran in recent years. The goal of the visit was to discuss acquisition of Iranian publications for the Middle East section of the Library. Also, he discussed Iranian plans for a new facility to house national archives and other library resources. During the visit, former Iranian President Mohammad Khatami called for bringing down the wall of mistrust (Wright, 2004).

In 2005, the Bam earthquake evoked sympathetic reactions around the world. The U.S. government promptly responded to the Iranian request for international assistance. The Iranian government arranged for entry into the country of American specialists in earthquake recovery. However, there were difficulties in both logistics and coordination. Eventually, the American team joined specialists from more than 40 other countries in assisting in the recovery operation (Garvelink, 2004).

While the immediate international response was impressive, the follow-up international meetings to consider longer term assistance were less successful. Tens of millions of foreign assistance dollars were pledged by many countries, including the United States. However, the amount of assistance that was actually delivered was a very small percentage of that promised.

In 2006, an unfortunate incident occurred that has left lasting scars on American and Iranian specialists interested in cooperation and particularly on Iranian professors at Sharif University of Technology. More than 40 faculty members of the university who received American visas to participate in the fourth Reunion of Alumni of Sharif University in California were denied entry into the United States at the San Francisco airport and other arrival points. Some were even placed in jail for a short time until they were able to arrange for return flights to Iran. The U.S. government explained that since the purpose of the reunion was to enhance the technological capabilities of Iran, the entry of the professors was not in the security interests of the United States (Sheikoleslami, 2006).

In the fall of 2006, former President Khatami visited a number of American organizations in several cities. Fortunately he included a meeting with leading American scientists and engineers in his itinerary. This recognition of the important role of the technical communities in the two countries in promoting understanding set an important framework for the discussion that we are having today.

In the fall of 2006, the Department of State launched its new visitor program for Iranians by inviting about 20 Iranian medical scientists to the United States for three weeks. All expenses are covered by the U.S. government for these programs, and the visa process is dramatically shortened. In subsequent months, the Department of State issued invitations to Iranian specialists in the fields of drug addiction, emergency response to natural disasters, and foodborne diseases.

This is the first program in recent years whereby the U.S. government pays expenses for visits by large groups of Iranians to the United States, and it represents a significant policy initiative. At the same time, these programs must be carefully designed and implemented to ensure that they are not misused simply for tourism. To this end, the programs should have high scientific standards both in the selection of the participants and in the design of the programs in the United States.[6]

In December 2006, the American Chemical Society informed the 36 Iranians who were members of the society that they could no longer retain their memberships due to legal problems. The specific issue is whether the society can provide member services to Iranians, including special prices on publications and meetings, access to databases, and training programs without a

[6] Information obtained from Office of Under Secretary for Public Diplomacy and Public Affairs, U.S. Department of State, September 15, 2007.

special license from the U.S. Department of the Treasury. The society has applied for a license from the Department of the Treasury, which we hope will soon be issued and would enable Iranians to rejoin the society. At the same time, we are concerned that this problem could affect Iranian members of other societies. There about 300 organizations in the United States that can be considered scientific societies (Bhattacharjee, 2007).

Our task is to ensure that this workshop becomes a beacon that lights the way for a surge in scientific contacts. We are experiencing too many negative events that constantly narrow the trail for scientific cooperation that could lead to better understanding at both the nongovernmental and governmental levels. For decades, Iran has been known as a land of engineers and doctors, while the United States has built its economy on the innovative skills of its scientists and engineers. Who can deny the value of joining efforts in a steady march to peace and prosperity?

DISCUSSION

Glenn Schweitzer: If a U.S. organization wants to teach short courses, the organization would have to obtain a license from the Department of Treasury. If we sign Iranians up and give short courses for them by Internet, it would be interpreted as a service, and a license would be required. In our role as a leadership organization we try hard to convince the government that we do everything possible to encourage the scientific community to obey the law. So when we are contacted, we render our judgment and urge the interested party to talk to the Department of the Treasury.

Yousef Sobouti: It is one of those cases that Professor Guyon was referring to when he said that certain rules should be broken.

William Wulf: We have a saying in English: "I would rather ask for forgiveness than for permission."

Schweitzer: Fortunately, we have a license to cooperate with Professor Sobouti and his colleagues in the general area addressed at this workshop for the next 28 months.

Sobouti: In this room there are many people from Shiraz University, from those days when the University of Pennsylvania was reshaping Shiraz University. Do you have recommendations for any of us?

Schweitzer: Yes; we need a Fulbright program.

REFERENCES

Amuzeger, J. 1966. Technical assistance in theory and practice, the case of Iran. New York, Washington, London: F.A. Praeger.

Bhattacharjee, Y. 2007. ACS drops Iranian members citing embargo. Science 315:1777.

Bill, J. A. 1988. The Eagle and the Lion: the Tragedy of American-Iranian Relations. New Haven: Yale University Press.

Garvelink, W. J. 2004. United States Response to the Earthquake in Bam, Iran, Special Report. Washington, D.C.: U.S. Agency for International Development.

Office of Public Affairs. 2004. Treasury Issues General License for Publishing Activities. Press release JS-2152. Washington, D.C.: U.S. Department of Treasury.

Sheikoleslami, Z. 2006. Letter to Secretary of State Condoleezza Rice.

The Rockefeller University. 1975. A report to the imperial government of Iran on the feasibility of establishing a new biomedical research institute. New York: The Rockefeller University.

Wright, R. November, 5 2004. Librarian of Congress touches many bases on Iran visit. The Washington Post, p.20. Available online at *www.washingtonpost.com/wp-dyn/articles/A26490-2004Nov4.html.*

13

Opportunities for International Collaboration in Earthquake System Science

THOMAS JORDAN
University of Southern California

Earthquakes and their effects pose the greatest natural threat to life and property in many urban regions throughout the world. Two prominent examples are Los Angeles, California, where I live and work, and Tehran, Iran, the host city for the international workshop on Science as a Gateway to Understanding. From my perspective as a geoscientist, these megacities are remarkably similar. Each is bounded by high mountains rising thousands of meters above fertile alluvial slopes and arid sedimentary plains. Their stunning but seismic geographies are actively shaped by folding and faulting in the boundary zones between gigantic tectonic plates.

Tehran and Los Angeles each comprise more than 12 million people; consequently, they account for much of their respective national total earthquake risk. Measured as annualized economic losses, almost one-half of the total earthquake risk for the United States comes from Southern California; of that, about 25 percent comes from the Los Angeles metropolitan area alone (FEMA, 2000). I am not aware of a comparable synoptic risk quantification for Iran, but hazard assessments and studies of building fragility suggest that Tehran's fraction of the national earthquake risk may be even higher (Tavakoli and Ashtiany, 1999; CEST-JICA, 2000; EMI, 2006; Jafari, 2007).

Megacity earthquakes can jeopardize prosperity and social welfare, and so it is in our common interest to know more about them and learn how to work together to reduce societal risks. Iran's long history provides a remarkable record of earthquake activity pertinent to this end (Ambraseys and Melville, 1982; Berberian, 1994). During the past 13 centuries, nine earthquakes with magnitudes greater than 7 have occurred less than 200 kilometers from Tehran. The last, in 1962, killed more than 12,000 people. Even much smaller, more frequent events can cause considerable damage. The magnitude-6.2 Firuzabad-Kojur earthquake, which struck a mountainous region 70 kilometers north of Tehran on May 28, 2004, killed 35 people, and preliminary assessments of its economic damage exceeded 125 billion rials.

As citizens of "earthquake country," many of us at this workshop share an interest in the earthquake problem. My focus will be on its scientific dimensions. Of course, engineering conditions are no less important. In particular, I will outline some of the key areas where scientific collaboration among Iran, the United States, and other countries might lead to new understanding of earthquake behavior that can help reduce risk. My discussion is intended to support a broader thesis: the potential for scientific cooperation to address our common environmental problems—water and energy supply, pollution, climate change, ecological degradation, as well as earthquakes—can be a strong force for developing crosscultural understanding and improving international relations.

SEISMIC RISK ANALYSIS

Earthquakes proceed as cascades in which the primary effects of faulting and ground shaking induce secondary effects, such as landslides, liquefaction, and tsunamis. They set off destructive processes within the built environment, such as fires and dam failures (NRC, 2003). *Seismic hazard* can be defined as a forecast of the intensity of these primary effects at a specified site on Earth's surface during a future interval of time.

In contrast, *seismic risk* is a forecast of the damage to society that will be caused by an earthquake, usually measured in terms of casualties and economic losses in a specified area. Risk depends on the hazard, but it is compounded by a community's *exposure*—its population and the extent and density of its built environment—as well as its *fragility*, the vulnerability of its built environment to seismic hazards. Risk is lowered by *resiliency,* or how quickly a community can recover from earthquake damage. The "risk equation" expresses these relationships in a compact (though simplistic) notation:

$$\mathit{risk} = \mathit{hazard} \times \mathit{exposure} \times \mathit{fragility} \div \mathit{resiliency}$$

Risk analysis seeks to quantify the risk equation in a framework that allows the impact of political policies and economic investments to be evaluated and thereby to inform the decision-making processes relevant to risk reduction.

Risk quantification is a difficult problem because it requires detailed knowledge of natural and built environments, as well as an understanding of both earthquake and human behaviors. Moreover, risk is a rapidly moving target, owing to the exponential rise in the urban exposure to seismic hazards. Calculating risk involves predictions of how civilization will continue to develop, which are highly uncertain. Not surprisingly, the best risk models are maintained by the insurance industry, where the losses and payoffs can be huge. However, the information from insurance risk models is usually proprietary and restricted to portfolios that represent (by design) a small fraction of the total exposure.

The synoptic risk studies needed for policy formulation are the responsibility of public agencies, and their accuracy and efficacy depends on technological resources not yet available in many seismically active regions. Risk assessments can be improved worldwide through international collaborations that share the expertise of earthquake scientists and engineers from countries with well-developed risk reduction programs. For example, many countries have benefited from the information about regional hazards produced by the Global Seismic Hazard Assessment Program during the United Nations International Decade for Natural Disaster Reduction (Giardini et al., 1999; Tavakoli and Ashtiany, 1999).

The first synoptic view of earthquake risk in the United States was published by the Federal Emergency Management Agency (FEMA) less than a decade ago (FEMA, 2000). This study obtained an annualized earthquake loss for California of $3.3 billion per year. However, it was based on a rather limited database of building stock and did not consider local site effects (e.g., soft soils) in computing the seismic hazard. A parallel but more detailed study by the California Division of Mines and Geology (now called the California Geological Survey) calculated a statewide expected value that was twice as large (CDMG, 2000). A revision of FEMA's 2000 report is currently underway using advanced methodologies and better inventories of buildings and lifelines.

Risk estimates have been published for California's historic earthquake events, such as the 1906 San Francisco earthquake (Kircher et al., 2006), and inferred from geologic data on the locations and magnitudes of prehistoric fault ruptures, such as the Puente Hills blind thrust system that runs beneath central Los Angeles (Field et al., 2005). The results are sobering. The ground shaking from a major earthquake on the Puente Hills Fault (magnitude 7.1-7.5), if it occurred during working hours, would probably kill 3,000 to 18,000 people and cause direct economic losses of $80 billion to $250 billion (Field et al., 2005). The large range in the loss estimates comes from two types of uncertainty: the natural variability assigned to the earthquake scenario (aleatory uncertainty) as well as our lack of knowledge about the true risks involved (epistemic uncertainty).

According to a similar scenario study, the loss of life caused by earthquakes of magnitude 6.7-7.1 on the North Tehran, Mosha, or Ray faults in greater Tehran ranges from 120,000 to 380,000 (CEST-JICA, 2000). The casualty figures for comparable earthquake scenarios in Los Angeles and Tehran thus show an order-of-magnitude difference, which derives primarily from the greater fragility of the built environment in Tehran. This comparison underlines the fact that the implementation of seismic safety engineering is the key to seismic risk reduction in urban areas.

STRATEGIES FOR SEISMIC RISK REDUCTION

I will illustrate the basic strategies for reducing seismic risk using California examples. The strategies can be categorized according to the four factors in the risk equation. For example, the exposure to hazard can be limited by land-use policies, such as the Natural Hazards Disclosure Act, passed by the California state legislature in 1998. The law requires that sellers of real property and their agents provide prospective buyers with a "natural hazard disclosure statement" when the property being sold lies near an active fault or within other state-mapped seismic hazard zones. This type of *caveat emptor* is typical of the weak compliance provisions in most land-use regulations. The high land values and population pressures in Los Angeles, where "sprawl has hit the wall," make the enactment of more stringent land-use policies quite difficult. We can thus expect seismic exposure to continue rising in proportion to urban expansion and densification.

A more effective strategy is to reduce the structural and non-structural fragility of buildings using building codes and other seismic safety regulations, performance-based design, and seismic retrofitting. The seismic safety provisions in the California building codes have been substantially improved by the tough lessons learned from historical earthquakes; in particular, revisions have corrected the design deficiencies identified in the aftermath of the destructive 1933 Long Beach, 1971 San Fernando, 1989 Loma Prieta, and 1994 Northridge earthquakes.

The efforts to promote seismic retrofitting have achieved mixed results. A 1981 Los Angeles city ordinance led to the demolition or retrofitting of almost its entire stock of unreinforced masonry buildings, the most fragile and dangerous class of inhabited structures. However, a state law regulating the seismic safety of hospitals, passed after the 1994 Northridge earthquake, has proven to be economically infeasible. Faced with the specter that many hospitals would be shut down rather than be retrofit, the legislature has postponed the compliance date for basic life-safety provisions of the law and is back-peddling on its long-term goal that all hospitals be capable of serving the public after earthquake disasters.

The latter requirement typifies performance-based design. Performance-based design goes beyond the building code requirements for life-safety by improving the ability of structures to retain a specified degree of functionality after episodes of seismic shaking (SEAOC, 1995). The impetus for performance-based design, largely economic, has raised new challenges for earthquake science and engineering (FEMA, 2006). In particular, engineers must be able to predict more accurately the damage state of structural systems—not just the system components—requiring more detailed descriptions of the ground motion. A full structural analysis uses complete time histories of ground motion to account for the nonlinearities in the structural response and in its coupling with near-surface soil layers. In California, the Pacific Earthquake Engineering Research (PEER) Center at Berkeley has organized a multi-institutional research program for advancing performance-based design.[1]

Community resiliency can be enhanced through better emergency response, insurance investments, catastrophe bonding, and state-funded recovery assistance. All of these tools are applicable to a wide range of natural and human hazards, including wildfires, severe storms, floods, epidemics, and terrorism. However, effective preparation and response to multiple hazards depends on a balanced view of relative risks. In the United States, there is concern that the recent emphasis on terrorist threats has distracted officials from efforts to prepare for natural disasters. The poor performance of the emergency response to Hurricane Katrina and subsequent disaster-recovery programs, especially in the hard-hit city of New Orleans, illustrate the need for better coordination

---

[1] See *peer.berkeley.edu/*.

and planning among local, state, and federal agencies (White House, 2006). One mechanism for improving coordination and planning is to conduct emergency response exercises based on realistic disaster scenarios.

Disaster mitigation can be enhanced by education. Public education is especially critical in preparing the response of megacities to catastrophic event cascades, during which government aid to the population might be insufficient and delayed (Perry et al., 2008). In the case of earthquakes, public awareness of the problem is greatly heightened after disruptive events, which motivate people to prepare for future disasters. Even small earthquakes, if widely felt, can provide "teachable moments," as can the anniversaries of famous disasters. In 2006, the centenary of the 1906 San Francisco earthquake motivated an extensive and successful public education campaign throughout California (USGS, 2006).

The first factor in the risk equation—the seismic hazard—is qualitatively different from the other three. We have no direct means to reduce the primary hazards of faulting and ground shaking. Earthquakes involve great forces of nature that will remain beyond human control for the foreseeable future. Nevertheless, the hazard level sets the risk, and the properly characterizing seismic hazard—forecasting earthquakes and their effects and charting earthquake cascades as they are happening—is therefore critical to risk reduction. For instance, current hazard forecasts contain large epistemic errors that compromise the effectiveness of risk analysis when guiding political policies and economic decisions. One role of earthquake system science is to reduce these uncertainties by improving our statistical and physical models of earthquake processes.

EARTHQUAKE SYSTME SCIENCE

A geosystem is a representation of nature defined by the terrestrial behavior it seeks to explain (NRC, 2000). In the case of an active fault system, the ground motion caused by a fault rupture is one of the most interesting behaviors from a practical perspective, because experience tells us that fault displacement and concomitant ground shaking are the primary seismic hazards for cities such as Tehran and Los Angeles. System-level hazard analysis can be exemplified by the following set of problems:

- Identify the active fault traces in a region to predict the maximum displacements that might occur across them.
- Predict the intensities everywhere in the region occupied by the network from the shaking intensities recorded on a sparse network of seismometers during an earthquake.
- Forecast the distribution of the shaking intensities in a region from all future earthquakes.

A basic methodology for solving the seismic forecasting problem is probabilistic seismic hazard analysis (PSHA). Originally developed by earthquake engineers, PSHA estimates the probability that the ground motions generated at a geographic site from all regional earthquakes will exceed some intensity measure during a time interval of interest, usually a few decades. A plot of the exceedance probability as a function of the intensity measure is called the *hazard curve* for the site. In downtown Los Angeles, for instance, typical estimates of the exceedance probabilities for peak ground acceleration (PGA)—a commonly used intensity measure—are 10 percent in 50 years for PGA $\geq 0.6g$ and 2 percent in 50 years for PGA $\geq 1.0g$, where g is the acceleration of gravity at Earth's surface (9.8 m/s^2). Other useful intensity measures are peak ground velocity (PGV) and the maximum spectral acceleration at a particular shaking frequency. From hazard curves, engineers can estimate the likelihood that buildings and other structures will be damaged by earthquakes during their expected lifetimes, and they can apply the performance-based design and seismic retrofitting to reduce structural fragility to levels appropriate for life-safety and operational requirements.

A seismic hazard map is a plot of the intensity measure as a function of site position for fixed exceedance probability. The official seismic hazard maps for the United States are produced by the National Seismic Hazard Mapping Project, managed by the U.S. Geological Survey. Seismic hazard maps are critical ingredients in regional risk analysis. For example, the FEMA (2000) and CDMG (2000) risk studies were based on the 1995 edition of the National Seismic Hazard Map (NSHMP, 1996). The revisions to the FEMA assessment are incorporating the better knowledge of seismic hazards encoded in the 2002 NSHMP edition. The latest edition, NSHMP (2008), has just been released and it will be used for the 2012 revisions to the Uniform Building Code.

The system-level study of earthquake hazards is "big science," requiring a top-down, interdisciplinary, multi-institutional approach. The Southern California Earthquake Center (SCEC) is funded by the U.S. National Science Foundation (NSF) and U.S. Geological Survey (USGS) with a mission to coordinate an extensive research program in earthquake system science. This program involves more than 600 experts at more than 62 research institutions (Jordan, 2006a).[2] Southern California's network of several hundred active faults forms a superb natural laboratory for the study of earthquake physics; its seismic, geodetic, and geologic data are among the best in the world. SCEC's mission is to use this information to develop a comprehensive, physics-based understanding of the Southern California fault system, and to communicate this understanding to society as useful knowledge for reducing seismic risk.

One of the goals of the SCEC program is to improve the techniques of PSHA through physics-based, system-level modeling. PSHA involves the manipulation of two types of subsystem probabilities: the probability for the occurrence of a distinct earthquake source during the time interval of interest, and the probability that the ground motions at a site will exceed some intensity measure conditional on that event having occurred. The first is obtained from an earthquake rupture forecast (ERF), whereas the second is computed from an attenuation relationship (AR), which

---

[2] See *www.scec.org*.

quantifies the distribution of ground motions as they attenuate with distance from the source.

The *ERF* that underlies the current U.S. national seismic map (NSHMP, 2008) is "time-independent" in that it assumes that earthquakes are random in time (Poisson distributed); in other words, it calculates the probabilities of future earthquakes ignoring any information about the occurrence dates of past earthquakes. However, owing to stress-mediated fault interactions and seismicity triggering, earthquakes are known not to be Poisson distributed. A major SCEC research objective is to develop time-dependent forecast models that include more information about the region's earthquake history. In the early 1990s, an SCEC-sponsored Working Group on California Earthquake Probabilities published a time-dependent *ERF* for Southern California (WGCEP, 1995). SCEC has more recently collaborated with the U.S. Geological Survey, and the California Geological Survey to produce the first comprehensive Uniform California Earthquake Rupture Forecast (WGCEP, 2007). The long-term (time-independent) model that underlies the UCERF was developed in partnership with the National Seismic Hazard Mapping Project, which has incorporated the results into its most recent release (NSHMP, 2008).

In the WGCEP forecasting models, the event probabilities are conditioned on the dates of previous earthquakes using stress-renewal models, in which probabilities drop immediately after a large earthquake releases tectonic stress on a fault and rise as the stress re-accumulates. Such models are motivated by the elastic rebound theory of the earthquake cycle and calibrated for variations in the cycle using historical and paleoseismic observations (WGCEP, 2003; Field, 2007b).

WGCEP (2007) estimates that, in the Los Angeles region, the mean 30-year probability of an earthquake with a magnitude equal to or greater than 6.7—the size of the destructive 1994 Northridge event—is about 67 percent. Because larger earthquakes occur less frequently, the chances of a magnitude ≥ 7.5 earthquake in the Los Angeles area during the next 30 years drop to about 18 percent. For the much larger Southern California region, the equivalent odds of a magnitude ≥ 7.5 event increase to 37 percent. The comparable value for Northern California is significantly less — about 15 percent — primarily because the last ruptures on the southern San Andreas fault in 1857 and circa 1680 were less recent than the 1906 rupture of the northern San Andreas fault. Sufficient stress has reaccumulated of the southern sections of the fault to make a large rupture more likely. The UCERF model will be used by decisionmakers concerned with land-use planning, the seismic safety provisions of building codes, disaster preparation and recovery, emergency response, and earthquake insurance; engineers who need estimates of maximum seismic intensities for the design of buildings, critical facilities, and lifelines; and organizations that promote public education for mitigating earthquake risk.

A second type of time-dependent ERF conditions the probabilities using seismic-triggering models calibrated to account for observed aftershock activity, such as epidemic-type aftershock sequence (ETAS) models (Ogata, 1988). In California, the Short-Term Earthquake Probability (STEP) model of Gerstenberger et al. (2005) has been turned into an operational forecast that is updated hourly.[3] The STEP forecast is a useful, though experimental, tool for aftershock prediction as well as the conditioning the long-term probabilities of large earthquakes on small events that are potential foreshocks. It should be emphasized, however, that the current probability gains in the latter application are relatively small.

The SCEC program seeks to improve time-dependent ERFs through better understanding of earthquake predictability. We have seen how long-term (decades to centuries) and short-term (hours to days) predictability are being exploited by operational time-dependent forecasting models. The challenge is to unify the forecasting models across the temporal scales, a task that requires a better understanding of intermediate-term (weeks to years) predictability. The research toward such unification is now focused on insights into the physical processes of stress evolution and seismic triggering (Toda et al., 2005). The SCEC-USGS Working Group

---

[3] See *pasadena.wr.usgs.gov/step*.

on Regional Earthquake Likelihood Models (RELM) is testing of a variety of intermediate-term models (Field, 2007a; Schorlemmer et al., 2007). Based on this experience, SCEC has formed an international partnership that is extending scientific earthquake prediction experiments to other fault systems through a global infrastructure for comparative testing. It is called the Collaboratory for the Study of Earthquake Predictability (Jordan, 2006b; CSEP, 2008). In the next section, I will elaborate on the exceptional opportunities presented by CSEP for international cooperation in earthquake system science.

Large earthquakes are rare events, and the strong-motion data from them are sparse. For this reason, a number of key phenomena are difficult to capture through a strictly empirical approach, including the amplification of ground motions in sedimentary basins, source directivity effects, and the variability caused by rupture-process complexity and three diemensional geologic structure. Therefore, a major objective of the SCEC program is to develop attenuation relationships that correctly model the physics of seismic wave propagation. Numerical simulations of ground motions play a vital role in this area of research, comparable to the situation in climate studies, where the largest, most complex general circulation models are being used to predict the hazards and risks of anthropogenic global change.

With NSF funding, SCEC has developed a cyber infrastructure for earthquake simulation, the Community Modeling Environment (CME), which allows scientists to construct system-level models of earthquake processes using high-performance computing facilities and advanced information technologies (Jordan and Maechling, 2003; Field et al., 2003). The CME infrastructure includes several computational platforms, each comprising the hardware, software, and scientific expertise (wetware) needed to execute and manage the results from different types of PSHA simulations. An example is the TeraShake platform for simulations of dynamic fault ruptures and ground motions on dense geographical grids. TeraShake simulations of ruptures on the southernmost San Andreas Fault have shown how the chain of sedimentary basins between San Bernardino and downtown Los Angeles form an effective waveguide that channels surface waves along the southern edge of the San Bernardino and San Gabriel Mountains (Olsen et al., 2006, 2008). SCEC is now increasing the performance of these computational platforms to take advantage of the petascale computational facilities that will be developed during the next several years. In the not-to-distant future, we will be able to incorporate much more physics into seismic hazard and risk analysis through system-level simulations.

INTERNATIONAL SCIENTIFIC PARTNERSHIPS

Earthquake system science relies on the premise that detailed studies of fault systems in different regions, such as Southern California, Japan, and Iran, can be synthesized into a generic understanding of earthquake phenomena. Achieving such a synthesis will depend on international partnerships that facilitate the development and comparison of well-calibrated regional models. I will briefly outline some of the salient opportunities opened by recent developments in earthquake system science.

EXPLORING THE EARTHQUAKE RECORD

The science of seismic hazard and risk is severely data-limited. Even in the most seismically active areas, the recurrence rates of large earthquakes are long compared to rates of urbanization and technological change. The last large earthquake on the southern San Andreas was in 1857, before the pueblo of Los Angeles became a city and before the pendulum seismometer was invented. According to WGCEP (2007), the 30-year probability of a large (magnitude ≥ 7.8) earthquake in Southern California is about 20 percent, too large for comfort, but small enough that it may be some time before we directly observe one or more of the "outer-scale" ruptures which dominate the behavior of the southern San Andreas system.

The power-law statistics of extreme events illustrate why progress in earthquake system science depends so heavily on comparative studies of active faults around the world. International scientific exchange has allowed much to be learned about continental faulting of the San Andreas type; e.g., from large strike-slip earthquakes that have occurred in Turkey, Tibet, and Alaska just during the last decade (Barka, 1999; Heaussler et al., 2004; Klingner et al., 2005). A plausible goal is the creation of an international database—a global reference library—for archiving the field and instrumental information recovered from such rare events.

A second obvious goal is to extend the seismicity catalogs for active fault systems backward in time. Countries like Iran with long historical records have a head start, but our knowledge of past activity can be significantly augmented using the new tools of paleoseismology and neotectonics to decipher the geologic record. Systematic paleoseismic investigations have elucidated a thousand-year history of San Andreas slip (Grant and Lettis, 2002; Weldon et al., 2005), and SCEC's current objective is to define slip rate and earthquake history of the southern San Andreas Fault system for the last 2000 years. Through international scientific exchange, these field-based techniques can be improved and applied to other fault systems.

The tectonics of Tehran and Los Angeles are both characterized by oblique convergence accommodated by complex systems of frontal thrust faults that are raising the Alborz Mountains and Transverse Ranges, respectively. A comparative study of these orogenic systems based on data from seismology, paleoseismology, remote sensing, and space geodesy would be a particularly good target for Iran-U.S. collaboration.

REAL-TIME SEISMIC INFORMATION SYSTEMS

A major advance in seismic monitoring and ground-motion recording is the integration of high-gain regional seismic networks with strong-motion recording networks to form comprehensive seismic information systems. A prime example of international collaboration is in the European-Mediterranean region, where the Network of Research Infrastructures for European Seismology (NERIES) is integrating more than 100 seismic monitoring systems and observatories in 46 countries into pan-European cyber infrastructure (Giardini, 2008).

On a regional scale, seismic information systems provide essential information for guiding the emergency response to earthquakes, especially in urban settings. Seismic data from a regional network can be processed immediately following an event and the results broadcast to users, such as emergency response agencies and responsible government officials, utility and transportation companies, and other commercial interests. The parameters include traditional estimates of origin time, hypocenter location, and magnitude, as well as Shake Maps of predicted ground motions conditioned on available strong-motion recordings, which can aid in damage assessments (Wald et al., 1999). In California, this type of information is provided by the California Integrated Seismic Network (CISN), which comprises more than a thousand seismic stations telemetered to central processing and data archiving facilities at the University of California, Berkeley, and the California Institute of Technology.[4]

Improvements in the real-time capabilities of these systems have opened the door to "earthquake early warning." EEW is the prediction of imminent seismic shaking at a set of target sites, obtained after a fault rupture initiates but in advance of the arrival of potentially damaging seismic waves. There are several EEW strategies (Kanamori, 2005), but the most common relies on a dense network of seismometers to transmit records of the first-

[4] See *www.cisn.org/*.

arriving (*P*) waves to a central processor that can locate the event, estimate its magnitude, and broadcast predictions to the target sites in near real time. In Southern California, the warning times in Los Angeles for earthquakes on the San Andreas Fault could be a minute or more, enough for individuals to prepare for shaking (e.g., by getting under a desk) and for certain types of automated decisions that might reduce damage and increase resiliency: slowing trains, stopping elevators, shutting gas lines, conditioning electrical grids, and so forth.

Several countries have already invested heavily in EEW systems. Japan's is the most advanced, but systems are also operational in Mexico, Taiwan, and Turkey (Horiuchi et al., 2005).[5] SCEC is participating with Berkeley and Caltech scientists in a USGS-sponsored project to test the performance of three EEW algorithms on the CISN system. However, the United States has been lagging in the development of EEW and could profit from more international involvement in this area.

DYNAMICAL MODELING

Numerical simulations of large earthquakes in well-studied seismically active areas are important tools for basic earthquake science because they provide a quantitative basis for comparing hypotheses about earthquake behavior with observations. Simulations are playing an increasingly crucial role in our understanding of regional earthquake hazard and risk because they can extend our knowledge to phenomena not yet observed. Moreover, they can also be used for the interpolation of recorded data in producing ShakeMaps and in the extrapolation of recorded data for earthquake early warning.

SCEC is applying simulation technology to the prediction of salient aspects of earthquake behavior, such as the influence of rupture directivity and basin effects on strong ground motions.

---

[5] See *www.jma.go.jp/jma/en/Activities/eew.html*.

Similar capabilities are being developed in Japan and Europe. Making this cyber infrastructure available for application in other regions is an excellent target for international scientific exchange. Such a program will entail the development of geologic models of regional fault networks and seismic velocity structures. Here, the SCEC experience in synthesizing three-dimensional structural representations may prove useful.

SEISMIC RISK ANALYSIS

From a practical point of view, the main role of earthquake system science is to promote risk reduction through better characterization of seismic hazards. For megacities like Tehran and Los Angeles, the key problem is holistic: how can we protect the societal infrastructure from extreme events that might "break the system," the way that Hurricane Katrina broke the city of New Orleans in 2005? Achieving this type of security depends on understanding how the accumulation of damage during an event cascade leads to urban-system failure. I will mention two ways that earthquake system science is contributing to this goal.

Earthquake simulations can provide cascade scenarios from which we can learn about, and possibly correct, the critical points of failure. In November 2008, the USGS will coordinate the Great Southern California ShakeOut, a week-long emergency-response exercise based on a SCEC simulation of a magnitude-7.8 rupture of the southern San Andreas Fault (Perry et al., 2008). ShakeOut will involve federal, state, and local emergency-response agencies, as well as several million citizens at schools and places of business. The objective of this disaster exercise is to improve public preparedness at all organizational levels.

SCEC is generating large suites of simulations that sample the likelihoods of future earthquakes. This capability for physics-based prediction of seismic shaking will someday replace empirical attenuation relationships in PSHA. It offers the possibility of an end-to-end ("rupture to rafters") analysis that embeds the built environment in a geologic structure to calculate more realistically earthquake risk for urban systems, not just individual structures.

The interests of basic and applied science converge at the system level. Predictive modeling of earthquake dynamics comprises a very difficult set of computational problems. Taken from end to end, the problem comprises the loading and eventual failure of tectonic faults, the generation and propagation of seismic waves, the response of surface sites, and—in its application to seismic risk—the damage caused by earthquakes to the built environment. This chain of physical processes involves a wide variety of interactions, some highly nonlinear and multiscale. Only through international collaboration can we extend such predictive models to all regions where the seismic risk is high.

EARTHQUAKE PREDICTION

Earthquake prediction *senso stricto*—the advance warning of the locations, times, and magnitudes of potentially destructive fault ruptures—is a great unsolved problem in physical science and, owing to its societal implications, one of the most controversial. Despite more than a century of research, no methodology can reliably predict potentially destructive earthquakes on time scales of a decade or less. Many scientists question whether such predictions will ever contribute significantly to risk reduction, even with substantial improvements in the ability to detect precursory signals; the chaotic nature of brittle deformation may simply preclude useful short-term predictions.

Nevertheless, global research on earthquake predictability is resurgent, motivated by better data from seismology, geodesy, and geology; new knowledge of the physics of earthquake ruptures; and a more comprehensive understanding of how active faults systems actually work. To understand earthquake predictability, scientists must be able to conduct prediction experiments under rigorous, controlled conditions and evaluate them using accepted criteria specified in advance. Retrospective prediction experiments, in which hypotheses are tested against data already available, have their place in calibrating prediction algorithms, but only true (prospective) prediction experiments are really adequate for testing predictability hypotheses.

The scientific controversies surrounding earthquake predictability are often rooted in poor experimental infrastructure, inconsistent data, and the lack of testing standards. Attempts have been made over the years to structure earthquake prediction research on an international scale. For example, the International Association of Seismology and Physics of the Earth's Interior convened a subcommission on Earthquake Prediction for almost two decades, which attempted to define standards for evaluating predictions. However, most observers would agree that our current capabilities for conducting scientific prediction experiments remain inadequate. Individual scientists and groups usually do not have the resources or expertise (or incentives) to conduct and evaluate long-term prediction experiments.

As a remedy, SCEC is working with its international partners to establish a Collaboratory for the Study of Earthquake Predictability. The goals of the CSEP project are to support scientific earthquake prediction experiments in a variety of tectonic environments; promote rigorous research on earthquake predictability through comparative testing of prediction hypotheses; and help the responsible government agencies assess the feasibility of earthquake prediction and the performance of proposed prediction algorithms. A shared, open-source cyberinfrastrcuture is being developed to implement and evaluate time-dependent seismic hazard models through comparative testing (CSEP, 2008). Testing centers have been established at SCEC, the Swiss Federal Institute of Technology in Zürich, and GNS Science in Wellington, New Zealand, and prediction experiments are now underway in several natural laboratories, including California, Italy, and New Zealand. Scientists from China, Japan, Greece, and Iceland have been participating in the development phase of CSEP, and we are encouraging other countries to initiate CSEP testing programs in the seismically active regions within their borders.

The research objectives of international partnerships in earthquake system science can be organized under four major goals: (1) discover the physics of fault failure and dynamic rupture; (2) improve earthquake forecasts by understanding fault-system evolution and the physical basis for earthquake predictability; (3) predict ground motions and their effects on the built environment by simulating earthquakes with realistic source characteristics and three-dimensional representations of geologic structures; and (4) improve the technologies that can reduce earthquake risk, provide earthquake early warning, and enhance emergency response. A common theme is the need to deploy cyberinfrastructure that can facilitate the creation and flow of information required to simulate and predict earthquake behaviors.

Toward this end, SCEC proposes the establishment of a Multinational Partnership for Research in Earthquake System Science (MPRESS) to sponsor comparative studies of active fault systems. The partnership would be organized to broaden the training of students and early-career scientists beyond a single discipline by exposing them to research problems that require an interdisciplinary, system-level approach and to enhance their understanding of how scientific research works in different countries, how different societies perceive the scientific enterprise, and how diverse cultures respond to scientific information about natural hazards.

This research was supported by the Southern California Earthquake Center. SCEC is funded by the NSF Cooperative Agreement EAR-0106924 and USGS Cooperative Agreement 02HQAG0008. The SCEC contribution number for this paper is 1210.

DISCUSSION

Thomas Jordan: You know, there are many stories about animal behavior before earthquakes. It is very easy to convince yourself that animals know what they are doing. After the big earthquake in Los Angles in 1971, I went out into the field to map the earthquake fault along the base of mountains. There were farms for raising horses. When I talked to the farmers, they said, "One hour before the earthquake the horses became very agitated." However, what they don't remember is one week before, when coyotes came down the mountains, the horses were also agitated. People tend to remember what happens before an earthquake, but not at other times. There is an historical record of earthquakes in Persia for more than two thousand years. Professor Ambrosias has looked at this. To properly interpret the data requires careful reading of the ancient texts and also geological investigations to try to match geologic features with ancient texts. It is a very important topic. It is a unique source of data. The historical record of earthquakes is extremely important to the study of earthquakes forecasting and prediction.

Yousef Sobouti: Do you have collaborations with institutions in neighboring countries, for instance Turkey?

Jordan: In Turkey we collaborate with four institutions.

Mostafa Damad: Is it possible to have that collaboration with Iranian institutions?

Jordan: Yes. Well I hope so. There are restrictions that have been imposed by the United States, but part of the reason we are here is to work with you to set up collaborations that make sense and then make sure that our governments understand what we are doing and approve. I see no reason why governments would not allow us to work on this common problem.

REFERENCES

Ambraseys, N. N. and C. P. Melville. 1982. A history of Persian earthquakes. Cambridge: Cambridge University Press.

Barka, A. A. 1999. The 17 August 1999 Izmit earthquake. Science 285: 1858-1859.

Berberian, M. 1994. Natural Hazards and the First Earthquake Catalogue of Iran. Tehran International Institute of Earthquake Engineering and Seismology p. 620.

California Division of Mines and Geology. 2000. An Evaluation of Future Earthquake Losses in California. Sacramento.

Center for Earthquake and Environmental Studies of Tehran and Japan International Cooperation Agency. 2000. The Study on Seismic Microzoning of the Greater Tehran Area in the Islamic Republic of Iran. Tehran.

Earthquakes and Megacities Initiative. 2006. Tehran Disaster Risk Management Profile. Megacities Disaster Risk Management Knowledge Base, Pacific Disaster Center.

Federal Emergency Management Agency. 2000. Report 366, HAZUS99 Estimated Annualized Earthquake Losses for the United States. Washington, D.C.: Federal Emergency Management Agency.

Federal Emergency Management Agency. 2006. Report 445, Next-Generation Performance-Based Seismic Design Guidelines. Washington, D.C.: Federal Emergency Management Agency.

Field, E. H. 2007a. Overview of the working group for the development of regional earthquake likelihood models (RELM). Seismol. Res. Lett. 78: 7-16.

Field, E. H. 2007b. A summary of previous working groups on California earthquake probabilities. Bull. Seismol. Soc. Am. 97: 1033-1053.

Field, E. H., H. A. Seligson, N. Gupta, V. Gupta, T. H. Jordan and K. Campbell. 2005. Probabilistic loss estimates for a Puente Hills blind-thrust earthquake in Los Angeles, California. Earthquake Spectra 21: 329-338.

Field, E. H., T. H. Jordan and C. A. Cornell. 2003. OpenSHA: a developing community modeling environment for seismic hazard analysis. Seismol. Res. Letters 74: 406—419.

Frankel, A., C. Mueller, T. Barnhard, D. Perkins, E. Leyendecker, N. Dickman, S. Hanson and M. Hopper. 1996. National seismic hazard mapping program, national seismic-hazard maps: documentation June 1996. Washington, D.C.: U.S. Geologic Survey.

Frankel, A., M. D. Petersen, C. S. Muller, K. M. Haller, R. L. Wheeler, E. V. Leyendecker, R. L. Wesson, S. C. Harmsen, C. H. Cramer, D. M. Perkins & K. S. Rukstales. 2002. National Seismic Hazard Mapping Program, Documentation for the 2002 Update of the National Seismic Hazard Maps. Washington, D.C.: U.S. Geologic Survey.

Gerstenberger, M., S. Wiemer, L.M. Jones & P.A. Reasenberg. 2005. Real-time forecast of tomorrow's earthquakes in California. Nature 435: 328-331.

Giardini, D., G. Grünthal, K. Shedlock & P. Zhang. 1999. The GSHAP global seismic hazard map. Ann. Geofisica 42: 1225-1230.

Giardini, D., T. van Eck, R. Bossu & S. Wiemer. 2008. Networking research infrastructures for earthquake seismology in Europe. Eos 89: 219.

Grant, L. B. and W. R. Lettis, editors. 2002. Paleoseismology of the San Andreas Fault System. Bull. Seismol. Soc. Am. 92(7).

Heaussler, P. J. and 10 others. 2004. Surface rupture and slip distribution of the Denali and Totschunda faults in the 3 November 2002 M 7.9 earthquake, Alaska. Bull. Seismol. Soc. Am. 9: S23-S52.

Horiuchi, S., H. Negishi, K. Abe, A. Kamimura and Y. Fujinawa. 2005. An automatic processing system for broadcasting earthquake alarms. Bull. Seismol. Soc. Am. 95:708-718.

Jafari, A.M. 2007. Time-independent seismic hazard analysis in Alborz and surrounding area. Natural Hazards 42(1):237-252.

Jordan, T. H. 2006a. Earthquake system science in Southern California. Bull. Earthquake Res. Inst. Tokyo 81: 211-219.

Jordan, T. H. 2006b. Earthquake predictability, brick by brick. Seismol. Res. Lett. 77: 3-6.

Jordan, T. H., and P. Maechling. 2003. The SCEC community modeling environment: an information infrastructure for system-level earthquake science. Seismol. Res. Lett. 74: 324-328.

Jordan, T. H., D. Schorlemmer, S. Wiemer, M. Gerstenberger, D. Jackson, P. J. Maechling, M. Liukis and J. D. Zechar. 2008. Collaboratory for the Study of Earthquake Predictability: Progress and Plans, Southern California Earthquake Center.

Kanamori, H. 2005. Real-time seismology and earthquake damage mitigation. Ann. Rev. Earth Planet. Sci. 33: 195-214.

Kircher, C. A., H. A. Seligson, J. Bouabid and G. C. Morrow. 2006. When the Big One strikes again: estimated losses due to a repeat of the 1906 San Francisco earthquake. Earthquake Spectra 22: S297-S339.

Klinger Y., X. Xu, P. Tapponnier, J. Van der Woerd, C. Laserre and G. King. 2005. High-resolution satellite imagery mapping of the surface rupture and slip distribution of the Mw ~7.8, November 14, 2001 Kokoxili earthquake (Kunlun fault, Northern Tibet, China). Bull. Seismol. Soc. Am. 95: 1970-1987.

National Research Council Committee on Basic Research Opportunities in Earth Science. 2001. Basic Research Opportunities in Earth Science. Washington, D.C.: National Academy Press.

National Research Council Committee on the Science of Earthquakes. 2003. Living on an Active Earth: Perspectives on Earthquake Science. Washington, D.C.: National Academy Press.

Ogata, Y. 1988. Statistical models for earthquake occurrence and residual analysis for point processes. J. Amer. Statist. Assoc. 83: 9-27.

Olsen, K. B., S. Day, J. B. Minster, Y. Cui, A. Chourasia, M. Faerman, R. Moore, P. Maechling and T. H. Jordan. 2006. Strong Shaking in Los Angeles Expected From Southern San Andreas Earthquake. Geophys. Res. Lett. 33: L07305.

Olsen, K. B., S. M. Day, J. B. Minster, Y. Cui, A Chouasia, R. Moore, P. Maechling and T. Jordan. 2008. TeraShake2: spontaneous rupture simulations of Mw 7.7 Earthquakes on the southern San Andreas Fault. Bull. Seismol. Soc. Am. 98: 1162 -1185.

Perry, S., D. Cox, L. Jones, R. Bernknopf, J. Goltz, K. Hudnut, D. Mileti, D. Ponti, M. Reichle, H. Seligson, K. Shoaf, J. Treiman and A. Wein. 2008. The ShakeOut Earthquake Scenario—a story that Southern Californians are writing. Washington, D.C.: U.S. Geologic Survey.

Petersen, M. D., A. D. Frankel, S. C. Harmsen, C. S. Mueller, Kathleen M. Haller, R L. Wheeler, R. L. Wesson, Y Zeng, O. S. Boyd, D. M. Perkins, N. Luco, E. H. Field, C. J. Wills and K. S. Rukstales. 2008. National Seismic Hazard Mapping Program, Documentation for the 2008 Update of the United States National Seismic Hazard Maps. Washington, D.C.: U.S. Geologic Survey.

Schorlemmer, D., M. C. Gerstenberger, S. Wiemer, D. D. Jackson, and D. A. Rhoades. 2007. Earthquake likelihood model testing. Seismol. Res. Lett. 78: 17-29.

Structural Engineers Association of California. 1995. Vision 2000: Performance-Based Seismic Engineering of Buildings. San Francisco: Structural Engineers Association of California.

Tavakoli, B. and M. Ghafory-Ashtiany. 1999. Seismic Hazard Assessment of Iran. Global Seismic Hazard Assessment Program.

Toda, S., R. S. Stein, K. Richards-Dinger and S. Bozkurt. 2005. Forecasting the evolution of seismicity in southern California: Animations built on earthquake stress transfer. J. Geophys. Res. 110: B05S16.

U. S. Geological Survey. 2006. 100th Anniversary of the 1906 Earthquake, People, Land and Water. Special issue 12(13).

Wald, D., V. Quitoriano, T. Heaton, H. Kanamori, C.W. Scrivner and C. B. Worden. 1999. TriNet "ShakeMaps": Rapid generation of instrumental ground motion and intensity maps for earthquakes in Southern California. Earthquake Spectra 15: 537-55.

Weldon II, R. J., T. E. Fumal, G. P. Biasi and K. M. Scharer. 2005. Past and future earthquakes on the San Andreas fault. Science 308: 966.

White House. 2006. The Federal Response to Hurricane Katrina: Lessons Learned. Washington D.C.: Office of the President.

Working Group on California Earthquake Probabilities. 1995. Seismic hazards in southern California: probable earthquakes, 1994-2024. Bull. Seismol. Soc. Am. 85: 379-439.

Working Group on California Earthquake Probabilities. 2003. Earthquake probabilities in the San Francisco Bay region: 2002-2031. U.S. Geologic Survey, Washington, D.C.

Working Group on California Earthquake Probabilities. 2007. Uniform California Earthquake Rupture Forecast, Version 2. Washington, D.C.: U.S. Geologic Survey.

LIST OF ABBREVIATIONS

AR—Attenuation Relationship
CDMG—California Division of Mines and Geology (now CGS)
CEST—Center for Earthquake and Environmental Studies of Tehran
CGS—California Geological Survey
CISN—California Integrated Seismic Network
CME—Community Modeling Environment
CSEP—Collaboratory for the Study of Earthquake Predictability
EEW—Earthquake Early Warning
ERF—Earthquake Rupture Forecast
ETAS—Epidemic Type Aftershock Sequence
FEMA—Federal Emergency Management Agency
JICA—Japan International Cooperation Agency
MPRESS—Multinational Partnership for Research in Earthquake System Science
NERIES—Network of Research Infrastructures for European Seismology
NSHMP—National Seismic Hazard Mapping Program
PEER—Pacific Earthquake Engineering Research Center
PGA—Peak Ground Acceleration
PGV—Peak Ground Velocity
PSHA—Probabilistic Seismic Hazard Analysis
RELM—Regional Earthquake Likelihood Models
SCEC—Southern California Earthquake Center
SEAOC—Structural Engineers Association of California
STEP—Short Term Earthquake Probability (model)
UCERF—Uniform California Earthquake Rupture Forecast
USGS—United States Geological Survey
WGCEP—Working Group on California Earthquake Probabilities

14

Science, Technology, and Innovation in the Knowledge Economy: Prospects for Cooperation

OUSMANE KANE
African Regional Centre for Technology

Scientific knowledge and its proper use have always been critical ingredients for economic performance and competitiveness. Today, the concept of a knowledge economy arising from the twin forces of globalization and technological progress, results in a closer linkage among science, technology, and innovation. Therefore, the knowledge economy requires proper knowledge management, which is a multidimensional process involving context, culture, content, mechanisms, infrastructure, and policy. It must address the dynamics of continuous change at the global, country, sector, and company levels. This raises many possibilities for enhancing growth and competitiveness by increasing productivity in all sectors of the economy and by adding value to local raw materials and natural resources.

In this regard, the knowledge economy has brought revolutionary changes to virtually all markets and sectors, but at the same time it also carries the risks of marginalization for countries, firms, or organizations that do not keep up with those rapid changes.

A successful knowledge economy requires a strong economic and institutional framework, a well-educated and skilled population, an efficient innovation system, and a dynamic information and communication infrastructure. Therefore, governments should lay the groundwork to enhance Africa's ability to seriously increase its scientific and technological potential, to resolutely embark on the knowledge economy, to promote sustainable development, and to become a key partner in global economy and trade.

The real wealth of any country is its people; and the core of any country's development process has always been knowledge, particularly in the fields of science and technology. In this regard, the fundamental facilitator of the knowledge economy is education, generally associated with a higher level of teaching and research. They are key factors for creating, sharing, disseminating, and effectively using knowledge for problem solving and innovation.

Africa's overall development, like in any region of the world, should be based on the tripartite elements—Economy-Energy-Environment—called the E3 method of sustainable development. But other methods are also important.

Currently, development in many advanced countries is mostly based on knowledge instead of raw materials and natural resources. Due to strong innovative systems, rapid advances in new and emerging technologies, such as information and communication technology (ICT), biotechnology, nanotechnology, and genomics are dramatically affecting all economic, social, administrative and cultural activities. The pervasive technological revolution resulting from these advances is now disrupting all kinds of relationships, transactions, and production systems of goods and services.

As a consequence of the ICT explosion that has led to worldwide interdependency and connectivity, globalization and competition have drastically increased and are leading to extensive shifts in world trade patterns and economic relations. Now, even corporate research and development are internationalized. Countries' or companies' competitiveness depends, more than ever, on their ability to access, adapt, utilize, and master scientific and technological knowledge for a continuous innovation process.

In this worldwide context, Africa should move away from its current position of passive technology spectator and urgently embark on a vigorous technology innovation strategy, at both the national and regional levels. Education and research are critical sectors where performance directly affects and even determines the quality and magnitude of Africa's development. They are the most important means we have at our disposal to develop human resources and impart appropriate skills, knowledge, and attitudes. As stated by the African Union in 2006

- education forms the basis for developing innovation, science, and technology in order to harness our resources, industrialize, and participate in the global knowledge economy. it is also the means by which Africa will entrench a culture of peace, gender equality, and positive African values
- research is critical for providing fundamental data on education in each country as well as essential information about instructional practice in school classrooms
- teacher education institutions should be engaged in research of a high order, as well as training teachers to do action research within their own teaching environments

Investing in human capital resources, research and development, and the promotion of innovation and entrepreneurship should be taken into due consideration through the establishment of trust and strong partnerships among all key stakeholders. These include government policy makers, higher education and research communities, production entrepreneurs, funding agencies, and consumer associations.

Although all dimensions, disciplines, and sectors are of great importance, we will mostly focus on science and technology issues. The new economic world order is mainly of a scientific and technological nature. However, the faculties of science and engineering at African universities register a small number of students —fewer than 30 percent of the total in many cases.

Therefore, we will consider the prospects of partnerships, in particular within South-South cooperation, in the context of science, technology, and innovation. We need to reverse the situation whereby Africa is the region that spends the smallest percentage of available funds on science and technology. As a consequence, the number of scientists and engineers is very small.

MAJOR CHALLENGES FOR AFRICA

Many resolutions have been adopted in the Organization of African Unity (OAU), now the African Union (AU), on the importance of science and technology as prime movers of the continent's socioeconomic development. Major continental initiatives such as the Lagos Plan of Action (Organisation of African Unity, 1980) and the New Partnership for Africa's Development[1] have made wide provisions for science and technology. However, there is still much to do for their implementation and translation into concrete programs of significant impact. As a consequence, Africa faces a number of paradoxes, being a continent with significant manpower and rich natural resources (water, minerals, petroleum, and biodiversity) but with the poorest people. It is confronted with several scourges such as unemployment, hunger, malnutrition, serious disease, lack of access to good education, poor leadership, crumbling infrastructure, and lack of energy and potable water. This situation is due in part to lack of strong innovative science and technology strategies with clear vision, firm commitment, and strong political will and leadership. These strategies should be endowed with the required resources, whether human, physical, or financial, and should be fully articulated in national socioeconomic development plans.

To break this vicious cycle and to become a region engaged in overall sustainable development and a respected partner in the global economy, Africa must establish an enabling environment

[1] See *www.nepad.org/*.

characterized by some fundamental parameters. Among those parameters are peace, a democratic and stable political system, good governance, social justice, and security of both people and goods together with a cautions application of scientific and technological achievements in the development process.

Also, the continent is confronted with many challenges, including the following:

- weak strategies for technology innovation and transfer
- Inadequate higher education and research systems with little innovative and inventive potential, a large brain drain (which must now be converted into brain gain), and the lack of national technological higher education, research, and innovation systems
- Lack of reliable data on scientific and technical potential (human resources, institutions, programs)
- Prevalence of micro-nationalism, resulting in rivalries instead of cooperation and integration based on comparative advantages
- Communication barriers (poor infrastructure, lack of telecommunications, languages, visa problems, cost of travels)
- Harmonization of a number of initiatives aiming at the promotion of science and technology throughout the continent

As stated by the AU in its Plan of Action for the Second Decade of Education for Africa (2006-2015) (African Union, 2006), Africa entered the Millennium with severe education challenges at every level. To cope with these challenges, conferences of the Ministers of Education have reiterated the need to increase access to education, improve quality and relevance, and ensure equity. Among specific challenges are the following:

- Lack of structural and organizational frameworks – institutions, infrastructures, extension and innovation mechanisms

- Lack of capacity and adequate resources – human, physical, financial, communicational, informational
- Africa is the region with the lowest expenditures devoted to higher education and research and, consequently, has the lowest number of scientists and engineers, with poorly equipped laboratories and packed lecture rooms;
- Adaptation to new world integrated higher education systems, such as the "license-master-doctorate" approach
- Inadequacy and fragmentation of curricula and research programs
- Partitioning and disarticulation of national socio-economic development plans and enterprises, particularly in the private sector
- Lack of motivation, leading to brain drain
- Lack of assessment of teachers, researchers, and program
- Lack of cooperation and partnerships among institutions, at national, regional, and international levels
- Poor management procedures and bureaucracy and frequet strikes
-
- Accordingly, the AU considers that the following priority areas should be addressed for the second decade of education:
-
- Improving supply and utilization of teachers
- Enhancing teacher competence
- Institutionalizing systematic career-long development of teachers
- Professionalizing and enhancing capacity for school leadership
- Improving teacher morale, working conditions, and social benefits
- Intensifying pedagogical research for continued improvement of teaching and learning

Scientific research and the entire process of technological innovation play a vital role in increasing economic performance, which in turn promotes employment, food security, access to energy, and wellbeing. They play a fundamental role in the improvement of the standard of living of a nation.

Sectors like agriculture and natural resource exploitation, which constitute the mainstay of the economy of the majority of countries in Africa, require the adoption and implementation of a strong strategy for technology innovation to address the numerous problems such as poverty, endemic diseases, hunger, and malnutrition that are drastically hindering Africa's sustainable development.

This strategy should be given the highest priority to allow for the creation of a greater range of sustainable wealth through long-lasting income and employment generation, resulting, for example, in more export opportunities for locally processed products. A relevant innovation system should be put in place with appropriate manpower, institutional framework, rules, and procedures, aiming at efficient and useful acquisition, master appropriation, dissemination and proper utilization of technological knowledge and packages.

The major elements of a strategy for technology innovation and transfer should be a specific component that is fully integrated into the national policy for science and technology. It has to be closely related to the national socioeconomic development policy and plan. It should be sustained by the three pillars of research, production, and market and composed of the following elements:

- Global needs assessment to clearly identify problems
- Vision for the future clearly describing the major development goals
- Strategic goals and general principles to set up well-defined objectives, with quantification, prioritization, and time-

frames based on the situation of the country in terms of food security, energy autonomy, and local raw materials

- Technology policy instruments as part of the structural and organizational framework to impart, conduct, and implement national strategies on science and technology for development. (Among the key bodies are the ministry of science and technology, the national council for science and technology, research-development institutions, centers for technology innovation and acquisition, intellectual property offices, and financing organizations for research and innovation.)
- Capacity building or strengthening of Africa's scientific and technical potential to address development problems which is crucial with regard to human resources development, infrastructures, and equipment for research and development
- Information and awareness to sensitize policy makers, economic operators, and civil societies as well as public opinion on scientific and technological achievements and their key role in development processes
- Cooperation and partnership to share experiences from success stories at the national, subregional, regional, and international levels in order to stimulate research activities and production, for example in agricultural or industrial sectors
- Assessment to measure and analyze achievements and gaps together with the incidence of key factors

Maintaining information and experience exchange is essential to stimulate research activity or industrial production. As a result of efficient and vibrant higher education and research systems, the establishment of technology innovation strategies in our countries, like in any other region of the world, could tap into global knowledge through trade, foreign investment, collaborative programs, and technology transfer within relevant channels.

African researchers or industrialists often do not have the appropriate scientific environment due to the lack of a well organized scientific and technological community including federations or associations organized by subject matter (physics, chemistry, biology) or by occupation (agronomists, nutritionists, electrical engineers). These researchers are generally isolated and affected by the absence, in their own countries, of valid interlocutors in their areas of specialization. In the same way, there is widespread inadequacy of scientific and technological information media (updated libraries and documentation centers, specialized periodicals, or publications) and lack of quick communication on new scientific and technological achievements (access to databases and data banks). Furthermore, the participation of African researchers in international scientific events like seminars, symposia, and congresses is most often hindered by the lack of financial means to cover registration, travel, and fees.

Moreover, the scarcity of scientists, technologists and engineers as well as of physical and financial resources makes it necessary to avoid duplication and to promote subregional, regional, and international cooperation. The cooperation and partnership strongly advocated by the Lagos Plan of Action (1980) as well as the New Partnership for Africa's Development (2001) could materialize through the establishment of consortia, joint programs, or thematic networks aimed at promoting higher education and research to pave the way for a strong technology innovation and transfer strategy. An international approach can harmonize national scientific and technological development as well as technological innovation strategies, including within the context of subregional political and economic organizations.

With regard to the need to support African technological higher education and research systems through dynamic inter-institutional partnerships, the following elements could be taken into consideration:

- Needs assessments of faculties of science, technology, and engineering
- Creation of a national strategy for science, technology, and innovation through interactive seminars involving higher education teachers and researchers as well as top policy makers

- Capacity building of technological higher education and research institutions, including exchanges of teachers and researchers through visiting programs, development of curricula on science policy, establishment of a framework for a national integrated system of higher education and research, and establishment of an extension and innovation unit with services to national or regional communities (private enterprises, public bodies, NGOs), and mechanisms for networking, connectivity, and sharing success experiences

The African Regional Centre for Technology (ARCT), an intergovernmental organization established in 1977 in Kaduna, Nigeria, under the aegis of the United Nations Economic Commission for Africa and the OAU, became operational in 1980 with its headquarters in Dakar, Senegal. According to its objectives, the ARCT aims to become an efficient tool in initiating, strengthening, coordinating, and integrating national, subregional, and regional technological capacities and strategies of African states.

Now, with the prospects and hope of the AU, the ARCT is fully committed to the implementation of Africa's Science and Technology Consolidated Plan of Action launched by the AU. The Centre, which is about to initiate a program on Managing Science, Technology, and Innovation for Africa's Sustainable Development in cooperation with various partners, is fully prepared to play a significant role in the development process, particularly with regard to issues related to technology innovation and transfer.

CONCLUSION

The capacity to generate, disseminate, and utilize scientific knowledge determines more and more the success of the participation of countries in the world economy. Being at the bottom in all areas of activity, the African population, in spite of the continent's richness in human and natural resources, obviously runs the risk of being abandoned and forgotten in an economically backward ghetto with restricted opportunities. The obstacles hampering technological innovation in Africa are numerous and stem from a set of economic, political, and structural parameters related to public authorities and to research institutions, development enterprises, and market specificities.

The phenomenon of universalization is increasingly subject to scientific and technological innovation intensified by, among other things, the spectacular progress in information and communication technology. Africa continues to be concerned by its mere survival, by the need to guarantee the daily subsistence of its population, to combat severe diseases, and by a number of other poverty issues. Therefore, for Africa to promote its sustainable development and cope with an increasingly competitive world, there is an urgent and real need to overcome these challenges and prepare for a vigorous technology innovation and transfer strategy. This strategy should be fully integrated into each country's science and technology policy, with a close relation to the national socioeconomic development plan.

To that effect, it is necessary to significantly increase the global resources allocated to technological higher education and research institutions to enable them to fully contribute to Africa's sustainable development and to define promptly for potential promoters all the technical and financial specifications and the practical modalities for the commercial exploitation of technological results at the industrial or craft level.

In this regard, strong partnerships involving other developing countries in the South (Asia and Latin America) as well as countries in Europe, North America, and Japan are of utmost importance.

The ARCT, with its wide experience and significant collaborative networks, is willing to assist African countries to formulate and implement realistic technology innovation and transfer strategies and to serve as a prime mover for the continent's sustainable development.

REFERENCES

African Union. 2006. Second Decade of Education for Africa (2006-2015): Plan of Action. Maputo, Mozambique: Conference of Ministers of Education of the African Union.

Organisation of African Unity. 1980. Lagos Plan of Action for the Economic Development of Africa 1980-2000. Addis Ababa, Ethiopia: United Nations Economic Commission for Africa.

15

Summary of Open Forum

YOUSEF SOBOUTI
Institute for Advanced Studies in Basic Science

Yousef Sobouti: This morning I requested that we write down very short summaries of the presentations. These short paragraphs will remind us of the issues and will be stimulants for this forum. I will invite comments and remarks from the floor.

SUMMARY OF FORMER PRESIDENT MOHAMMAD KHATAMI'S PRESENTATION: WHERE DOES SCIENCE GO?

Cultural, scientific, technological, and other intellectual relations between the nations of the east and west are not symmetrical and are polarized. This hinders the process of understanding.

Comments

Glenn Schweitzer: He commented that scientists are simply one extension of society. So when you talk about the role of scientists, you have to talk about the role of society. His point was that ethics are important for the entire society and that ethics will spread through scientists.

His second issue had to do with violence supported on the basis of religion. Violence that is justified on the basis of religion—Islam, Christianity, or any other religion—is very bad. The attitude of some leaders all over the world that he cited was, "If you are not with us, we can do what we want to you." His final point was that you have to replace violence with love. That was his philosophical point.

In the third issue on U.S.-Iranian relations, he made the point that the problem is much bigger than bilateral relationships and must be viewed within the entire global system which has developed. He characterized the systemic problem as the developed versus the developing world. This was the same point that was made in the newspaper this morning. He argued, understandably, that dialogue is the solution. But it must be based on fairness, equality, and justice. Whether you agree with that or not, it is important that we try to capture this idea in the proceedings.

Ferenc Szidarovszky: President Khatami made the important point, "If you don't have security in one place, you don't have security anywhere."

Sobouti: Yes, this point is to be emphasized.

SUMMARY OF WILLIAM WULF'S PRESENTATION: THE INNOVATION ECOLOGY

Solving many of the world's problems—such as climate change—will require innovation of new technologies. Unfortunately the ecology of laws, regulations, and institutions that support innovation were invented for technologies of the past and do not support current and future technologies well. We need an international process to rethink the elements of this ecology.

Comments

Etienne Guyon: I was impressed by the associations that Professor Wulf described among the Academy of Engineering, the Academy of Sciences, and the Institute of Medicine by saying that we should involve all of them in a given process. I was struck by his "Pasteur's Quadrant," saying that science and technology in developed and developing countries have to work hand-in-hand and not as separate entities in order to achieve a better world.

SUMMARY OF MOSTAFA MOHAGHEH DAMAD'S PRESENTATION: WISDOM: THE BEST GATEWAY TO UNDERSTANDING

I would like to make a distinction between the two words *science* and *wisdom*. I interpret *hikma* in Arabic to mean *wisdom*, but I am not sure whether it is used in the same sense in the western world. The best gateway to understanding is *hikma*, not science. Science, devoid of spirituality, sometimes enhances man's arrogance and ego. On the other hand, *hikma* encompasses both ethics and education.

Comments

Schweitzer: I had difficulty understanding Professor Damad's presentation on wisdom, which was very important. So I had a private conversation with him, and he explained that what he meant by wisdom includes four aspects:

- Spirituality
- Ethics
- Education
- Understanding the relationship between humans and the universe, which includes science

Mehdi Bahadori: I suppose a wise man is a person who has wisdom. In Christian epics, you have the birth of Jesus and the three wise men from the east who came to him. They followed the stars to Bethlehem. What is meant by wise men here? Are they scientists? Who are they?

Norman Neureiter: To me they were the three leaders of those countries. Sometimes, they are called the Three Kings from the Orient. They are translated as kings, not just wise men. So you think of them as wise leaders, not necessarily scientists, with no analysis of the word *wisdom* or the adjective *wise.*

Sobouti: The words *science* and *scientist* are new as used in the past 200 or 300 years. Two thousand years ago, there were no such words. So the three wise men of the Bible should be considered in the sense that Professor Damad discussed. After all, *hikma* and *wisdom* are very old notions. *Hakim* was the one who knew almost everything. This, at least, is the definition of *hakim* in Islamic societies.

Bahadori: In our literature we have used the word *hakim* to address a poet, Hakim Ferdousi, a mathematician, Hakim Khayyam, or a philosopher, etc. If Professor Damad were here, he would say they were wise men. But Ferdousi is a poet, not a scientist. Khayyam is a scientist and a poet at the same time.

Oliya: *Hakim*, in my opinion, is a holistic person who harmonizes his thoughts and actions with the laws of nature and the whole universe.

Sobouti: Yesterday, I had my differences with Professor Damad. A newborn baby has the potential to become wise and to acquire wisdom, but is not a wise person and doesn't possess wisdom at birth. Wisdom comes through later experiences and exposure to knowledge. In this respect, science is the most determining factor to enable one to acquire wisdom.

To conclude my point, let me quote Molana Rumi. The man, whose eight hundredth birthday the world is celebrating this year, is a Sufi, a poet, a theologian, and a philosopher. He is whatever you may wish to attribute to a *hakim*. In one of his poems he says, "If you are wearing blue spectacles, you will see the world blue." In all fairness, I should confess that I'm wearing the spectacles of science and see everything from the viewpoint of science. Professor Damad is wearing the spectacles of wisdom and sees the world from the viewpoint wisdom.

SUMMARY OF ABULHASSAN VAFAI'S PRESENTATION: INTERACADEMY COOPERATION: AN APPROACH TO UNDERSTANDING

While the basic concepts of science have not altered, social needs have changed dramatically. Science must address complex issues that are global in scale and must deal with difficult problems that can only be met by joint efforts. Interacademy collaboration based on the experience gained in the last ten years between Iranian and U.S. academies is one of the best ways to achieve understanding.

Comments

Abulhassan Vafai: Some of our French friends asked why I didn't address more comprehensive bilateral collaboration. At Sharif University of Technology, we have collaborations with countries all over the world. But in this workshop, I was merely presenting a case study, specifically the past experience of the United States and Iran.

Guyon: As far as my own country is concerned, many initiatives could be sought within the European Union.

Sobouti: Thank you, Professor Guyon. Tomorrow you are coming to Zanjan to visit the Institute for Advanced Studies in Basic Science (IASBS). There I will hand you a long list of collaborations that IASBS has had with French institutions.

SUMMARY OF YOUSEF SOBOUTI'S PRESENTATION: UNDERSTANDING OTHERS, THE SCIENCE WAY

The logic and methodology of exact sciences are universal and free of cultures, of beliefs, and of any manmade conventions. At the same time, science is the most vigorous driving force behind the development of all societies. A practice to use the logic and methodology of science in other areas of man's activities should, in principle, help people to better understand each other.

As to the inexact sciences, the world has not forgotten the two opposing economic schools of 10 to 15 years ago. The opposing factions were ready to annihilate each other because the principles that one side was upholding were not acceptable to the other side. Similarly, as to the issues of beliefs and religions, a person cannot possibly convince another person with different beliefs that he is right and the other is wrong.

Comments

Schweitzer: You said that the exact sciences can be used everywhere. I wish your statement were true. I think the more accurate statement would be that they should be available to be used everywhere. But there are many places where no capability exists to use the sciences, particularly in developing countries.

Masoumi Hamedani: I am afraid I didn't understand your essential point. Do you believe that other branches of human learning or human activity can reach someday the situation in which physics, for example, is found today?

Sobouti: My optimistic answer is yes. If a branch of knowledge acquires the universal and convention-free logic of physics, it will become quarrel-free as well. I find support of this belief in the history of science.

Guyon: I am often embarrassed when I talk about the exact sciences in the sense that our experimental sciences are not exact. There are fluctuations and errors. There is also the question of reproducibility, and fluctuations are not reproducible. Experimental science aside, there are many other domains that are as important as the exact sciences but have a totally different nature—history and sociology, for example. They are not reproducible. You will never again have the Second World War, or you will never know that if the murder in Sarajevo had not happened, the First World War would not have started in the same way. So history will never be an exact science. It cannot be because it cannot have the reproducibility of the experimental sciences.

Sobouti: I agree that history is not and cannot be an exact science.

Guyon: Yet it is highly valuable and indispensable.

Schweitzer: We have not given enough credit to the social and economic sciences. It is in the application phase that the social and economic sciences are important, but I do not think that you can equate the social and economic sciences to the natural sciences.

Sobouti: I thank you all for the active participation in the discussion. The points of view that you have offered are most valuable and will certainly enrich the proceedings.

SUMMARY OF ETIENNE GUYON'S PRESENTATION: LOVING AND SHARING SCIENCE: PIERRE-GILLES DE GENNES

I presented some of the elements of the tool box of one of the greatest physicists of the twentieth century who combined vigor with intuition and fantasy, open-mindedness with curiosity, and sharing with listening to others. His way of presenting science was based on simplicity. Images and analogies were always present, and pedantry and selfishness were excluded. He was a lover of life, as he was a lover of science. The model he offered to us opens the doors to understanding to a large class of people and not just scientists.

Comments

Guyon: To open a gateway to understanding, one needs keys to it. Pierre-Gilles de Gennes had this metaphor. A scientist is like a man in front of a locked door with many keys in his hand. A prudent person would ordinarily try to study the shape of the keys and the keyhole before trying a key. "But what I did instead," said de Gennes, "was to use the first key that I put my hand on; and much to my delight it opened the door. But then I realized that all the keys also opened the door." The important aspect is to dare to use a key.

Mandana Farhadian: I am familiar with Professor de Gennes' thoughts and research. Once he mentioned that for students who come from developing countries to the West, it is better that they choose a topic they can pursue in their home countries as well. Otherwise they might be compelled to stay in the foreign country with little benefit to their homeland. I believe this is very good advice to follow when considering exchanges of students or interuniversity collaborations.

Sobouti: One of the impressive qualities of Pierre-Gilles de Gennes was his search for physics around himself rather than in books. For him, physics was everywhere, in the flow of water from a tap, in the coiling of honey from a spoon, in cosmetics, and in everything and everywhere.

Bahadori: Newton was like that.

Sobouti: Pierre-Gilles de Gennes was not the only person to have this quality, but he was one of the few in contemporary times that we have known.

SUMMARY OF NORMAN NEUREITER'S PRESENTATION: SUCCESSES IN BUILDING INTERNATIONAL BRIDGES THROUGH SCIENCE

International scientific and technical cooperation can be a very useful instrument of an active and constructive foreign policy. It can also be useful in improving U.S. relations with Iran.

Comments

Sobouti: I don't think anyone disagrees with this wise statement.

Ousmane Kane: Increased relations between northern and southern countries should be emphasized.

Neureiter: Clearly, we can generalize it. A broadly informed policy is a useful instrument and can be very appropriate to countries in Africa and other areas. The context for the Americans is the relationship with Iran. But I think we should agree with Dr. Kane to generalize the statement.

SUMMARY OF AKIO MATSUMOTO'S PRESENTATION: BRIDGE FOR MUTUAL UNDERSTANDING

Economics is a branch of the social sciences. However, it has interdisciplinary marriages with mathematics, physics, and engineering and is expanding its scope. Dividing science into the social sciences and the natural sciences seems to be superfluous.

Comments

Schweitzer: There are fundamental differences between the natural and social sciences in terms of the evidence on which you base conclusions. It is a mistake to suggest the social sciences are just as much based on evidence as physics, for example; and so I have an objection in combining the natural and social sciences.

However, I think there should be a counterpoint that it is not a generally accepted position.

Guyon: Yesterday, I gave a word of warning that one cannot apply concepts devised for physics and mathematics to the social sciences in a straightforward way, for each field has its own complexities, different from those of others.

Sobouti: In my own presentation yesterday, I made a distinction between the exact sciences and the empirical ones. It is true that economics, sociology, the art of governance, and so have their rules and regulations. But those rules and regulations are taken from everyday practices. They are at most empirical and not as exact as the laws of nature that are valid everywhere and at all times and are free from any belief and culture. So I am also inclined to disagree with Akio.

Szidarovszky: I also disagree from the mathematical point of view. He tried to identify the social sciences as a portion of the natural sciences that is mathematical.

SUMMARY OF HOSSEIN MASOUMI HAMEDANI'S PRESENTATION: THE UNIVERSALITY OF SCIENCE: EXAMPLES FROM HISTORY

The universality of science cannot be guaranteed by its epistemological status alone. A truly universal science, while accessible to every man and woman, should also be conscious of its roots in different cultures and the overall characteristics of that culture's history.

Comments

Hyadi Khajehpour: I believe that science is a social institution. When we meet and talk about the relations between different scientific organizations and understanding of science, we are not talking about science in its proper sense, that is, laws, experiments, and related items. But we are talking mostly about the institutions that scientists have built, such as academies of sciences, universities, and laboratories. They are not only places where scientific work is carried out but are also institutions concerned with political problems. So I would agree with Dr. Masoumi. It is not the epistemological part of science that is important and brings understanding, but it is its social institutions that bring people together and make them talk about the things they have discovered or the things they plan to do. The scientific institutions are key mechanisms in bringing about understanding among different people.

Sobouti: You make a distinction between the epistemological part of the science and the institutions that are supporting and patronizing science. Your point is well taken. Academies of sciences around the world, science societies, and other organizations support science. But they are not directly involved with its epistemology. They in and of themselves help bring better understanding, yet they are not the laws of Newton.

Schweitzer: You have a point. But I think that physics institutions working together are not the same as social science institutions working together. There are so many judgments in social sciences that are tied up with politics that it really is not fair to compare them to the universality of physics. When we say science is universal, I think the natural sciences or the exact sciences are universal. But I am hesitant to say the social sciences are universal because they are laden with judgments that are hardly based on evidence.

Christian Duhamel: I am Christian Duhamel from the French Embassy, a scientific attaché and a mathematician in civil life. Cultivation of science should follow a certain philosophy: never leave a question of a child unanswered. Just an example: my daughter once came home and said she asked her chemistry teacher, "How did they find Avogadro's number?" The response was, "You didn't need to know this to do the exercises you are given." Such an attitude destroys the natural curiosity of the child. Can one really become a creative chemist without knowing how

Avogadro's number was conceived? This attitude, in my opinion, is prevalent not only in science education but also in other areas. You do not need to know that to do this problem, and don't ask further questions. You do not need to know that to live, so keep quiet. In this way, we prepare the child to live in a very technical world, as it is now, in blind serenity. He or she doesn't need to know the workings of the telephone, the Internet, the television, or the airplane in order to be able to use them in everyday life.

With such an approach, science can be a part of dictatorship. If we train children to submit to certain rules and use them to solve the problems assigned to them, we are depriving them from the freedom of thinking. If we want to democratize education, we must completely renovate the whole system of education of children.

Sobouti: Let's see if we can conclude from these lively discussions that educating people to become scientists is the task of scientific institutions, different from the epistemological aspects of science.

Guyon: I don't fully agree. We need scientists, but we need responsible citizens as well. We want to have science education, not just to create a good group of good scientists, but to enable people, everybody if possible, to have the basic understanding of the scientific methodology and the scientific facts. I would like to broaden the word *scientist*. The purpose is not to train just scientists but citizens, responsible citizens, with a science culture.

Sobouti: I agree with you.

SUMMARY OF MICHAEL CLEGG'S PRESENTATION: THE ROLE OF INTERNATIONAL SCIENTIFIC ORGANIZATIONS

Why should science emphasize international cooperation? All people of the world face common challenges such as global warming, water resource scarcity, and food security. Science can offer an optimal approach to mitigating these problems. To reach decision makers, several global organizations have evolved. One is the InterAcademy Panel (IAP), a global network of science academies, with the goal of providing decisionmakers with rational solutions to global challenges. The IAP works with an engineering analog, a medical analog, and a body that provides in-depth recommendations on global issues to policy makers.

Comments

Guyon: Clearly, academies are playing important roles. But in my opinion, their functions should not replace the responsibilities of their member citizens. On problems of ethics, for example, I have seen a number of ethics committees. I presume that they are knowledgeable and should do their jobs. On the other hand, such institutions should not prevent the citizens from being informed adequately to contribute.

SUMMARY OF BERNARD MAITTE'S PRESENTATION: SCIENCE AND CULTURE

Contemporary science has modified societies deeply. Because of its disciplinary character, however, it is often separated from culture. Past technologies that were developed through means different from modern techniques were intimately connected with the ways of life of the people. The question remains open whether one is justified to use any technology that science produces.

Comments

Guyon: Nanoscience, high energy physics, and related developments are getting closer and closer to becoming technologies. For example, we are now working with Bose-Einstein condensation. Maybe in 10 years, Bose-Einstein condensation—now a pure science—will become a technology as lasers did in the 1960s and 1970s. At the same time, it is important to know how new sciences like nanoscience will impact technology. This will give people more freedom of choice to accept or reject scientific achievements.

Khajehpour: I have a bit harsher understanding of Dr. Maitte's contribution. I think what he wants to say is that the modern sciences started by claiming to conquer and even, as Hume puts it, to rape nature. This was the way people thought and expressed their endeavors. Until the twentieth century the human being was somehow excluded from what was to be conQuéréd. But by the mid-century and onwards, with the very rapid advance of science, this conquering has included the human being himself. Science and technology do whatever is possible—to some extent irrespective of the real needs of the human being. We may not agree, but it is a fact that this "new human being" is quite a different subject. Now we are living with a different approach to science that I believe should be corrected and should be examined carefully.

Junes Ipakschi: In my opinion, there is no clear separation between science and technology. The technology of producing chemicals is part of science, as is medicine.

Bahadori: Have you heard the statement, "The scientist discovers what there is, but the engineer is the one who creates what there isn't?" Engineers utilize the basic knowledge available to them but create things that don't exist. In the case of medicine, the person who makes it is an engineer. He uses the basic knowledge that exists.

SUMMARY OF GLENN SCHWEITZER'S PRESENTATION: A HALF CENTURY OF SUCCESSES AND PROBLEMS IN U.S.-IRANIAN COOPERATION IN SCIENCE, ENGINEERING, AND MEDICINE

An example of past cooperative programs, in this case U.S. and Iran cooperation, can provide useful lessons that are helpful in designing and carrying out future programs. We need to learn from the past to inform the future.

Comments

Masoumi: Before the revolution, there was very strong scientific cooperation between Iran and the United States After the revolution there was a new phase with drastically reduced communications. Cooperation has always been dependent on political situations. I have three questions: How does this come about? Can we envisage another situation in which political changes do not affect scientific relations to such a degree? Has the new phase of cooperation between Iranian and American academies led to any concrete research projects or cooperation on specific problem?

Sobouti: You are saying that international collaboration depends on the support of governments and that individual roles can be enhanced or diminished, depending on the policies at higher levels. In any event, collaborations always need patrons. Don't we agree on that?

Schweitzer: You are certainly right. But there is one point that is important. Scientific collaboration can have the capability of withstanding political assaults, if you like. So in the case of Iran and the United States, the fact is that many Iranian students went to the United States, for example, and came back despite political changes in Tehran. They are still trying to maintain their collaboration, in spite of the political situation. The point is that cooperation in science has a built-in sustainability that is very useful in withstanding political difficulties and in surviving through political turmoil. Are there any success stories in recent times? I think the number of joint publications between Iranian and American scientists is very high relative to the number of joint publications between Iranians and scientists of other countries. In fact it's at the top of the list. The fact that there are so many joint publications, even at this most difficult time, supports the idea that cooperation can withstand political assaults.

Guyon: It would be very valuable if historians studied cases similar to U.S.—Iran collaborations. To wit is the case of the Soviet Union and the rest of the world during the Cold War era. Collaborations between scientists, and particularly space scientists, all over the world were excellent. I am very familiar with the difficulties of the peoples of Argentina and Chile with the rest of the world and also the case of French scientists during the difficult time of fascism in France. Despite the difficulties with the governments, scientific contacts were maintained without interruptions. I have followed very closely Chile after Pinochet, and I can say that there was no break in collaborations despite political difficulties.

SUMMARY OF THOMAS JORDAN'S PRESENTATION: OPPORTUNITIES FOR INTERNATIONAL COLLABORATION IN EARTHQUAKE SYSTEM SCIENCE

The study of earthquakes is a system science that requires international collaboration in order to sample earthquake behaviors in different tectonic environments. Opportunities exist for international collaboration in (1) prediction of strong ground motions, (2) earthquake rupture forecasting, and (3) education of the public about earthquake risk.

Comments

Khajehpour: The type of work that Thomas Jordan mentions is one of the best examples of collaboration that can be developed among different nations. It can withstand political upheavals in developing countries. This sort of cooperation is hard to develop, but when it is developed it will endure changes that may happen between the states. It will endure because it is scientific work and because determined scientists are involved. They come to a much wider understanding of what the scientific work is about.

Duhamel: I am in charge of the scientific collaboration between France and Iran. We have a lot of cooperation between France and Iran. It could be useful not only to have bilateral collaborations but also to think of multilateral and international cooperation. Presently we have collaborations with three universities in Iran: Sharif University, Tehran University, and Amir Kabir University. One of our projects is to create a common doctoral program in mathematics jointly among France, Spain, and probably Germany. Other countries may also join. Seismology is particularly of interest. Geologists are interested in data coming from earthquake-prone Iran.

Vafai: On behalf of Sharif University, I would like to emphasize what Dr. Duhamel just said. In the past 10-15 years, we have trained about 35 scientists in collaboration with France. The program started as a pilot project. The program required a student to spend a year in France to complete his thesis project. We began with 2-3 students and gradually increased the number to 10—15. Most of these scientists, with the exception of a few, are teaching in Iranian universities all over the country.

SUMMARY OF OUSMANE KANE'S PRESENTATION: SCIENCE, TECHNOLOGY, AND INNOVATION IN THE KNOWLEDGE ECONOMY: PROSPECTS FOR COOPERATION

Science, technology, and innovation have an important role to play for Africa's sustainable development. To that end, strong international partnerships through institutional collaborative programs are most welcome, particularly for capacity building.

Comments

Ipakschi: I am an engineer, and I think that science and technology are different.

Guyon: In fact, the difference is huge because of the lack of technological training of young people. This lack of technological training is related to the lack of equipment, not necessarily heavy equipment. Technology in most of Africa is at a low level. It is not uncommon to see very good African scientists. But these young scientists are not often capable of developing experiments. One should emphasize adequate basic training in experimental science.

Sobouti: My point is that developing a society is a very complex task because societies are very complex systems. In complex problems, one cannot pinpoint all impeding factors and make adequate provisions for them. At most, one can come up with an average and approximate solution. The example of the "train" that you mentioned is not satisfying: A train is a simple system. It either moves forward or backward, but a society is different. To develop a society, you need all possibilities at your disposal, and these possibilities include the talent of everyone that you may find in the society. To make those talents flourish, you need a high degree of literacy. It is good literacy that I wish to emphasize.

SUMMARY OF WILLIAM COLGLAZIER'S PRESENTATION: GLOBAL ENERGY CHALLENGES

Global energy challenges can be a source of friction and conflict, even war, between countries. If applied wisely, science and technology can help to find pathways to a secure energy future for all people. The objective should be to improve lives everywhere and to protect the environment.

Comments

Schweitzer: I would emphasize Iran's unique position in the world. Iran has the second largest liquid oil reserve in the world and the second largest gas reserve. Given that, we should acknowledge that cooperation in the energy area is very important for Iran. What do you think about the future of nuclear power and CO_2?

Colglazier: Nuclear power is not totally free of CO_2 because you have to manufacture and produce the equipment for the power plants. There are other potential environmental issues. So nuclear technologies are not totally benign. But the carbon emissions from nuclear power are very low compared to the other alternative energy sources.

Neureiter: The waste storage problem is not trivial.

Colglazier: Waste is a serious problem, but there are technical solutions. The political problems may be even greater than the technical problems, since not many populated areas are interested in having a nuclear waste storage site in their neighborhoods.

Sobouti: For carbon sequestration, what are the prospects at present?

Colglazier: There is a long way to go. It is going to be expensive, and much of the research and development is needed. The United States is just starting with some demonstration projects in three areas of the country, but it is not a near-term technology at the moment. Probably the most important thing we can do in terms of dealing with energy usage in transportation in the near term is being more efficient. There are many studies on technologies that could improve energy efficiency in automobiles significantly. In the United States, some of the big automobile manufacturers are not particularly eager to be forced to employ new technologies. If you are interested in relevant reports, the website of the U.S. National Academies is *www.nationalacademies.org*. If you are from a developing country, all our reports are available free of charge in PDF format.

SUMMARY OF FERENC SZIDAROVSZKY'S PRESENTATION: MY PERSONAL EXPERIENCE WORKING WITH PEOPLE OF DIFFERENT CULTURES AND BACKGROUNDS

I reported from my life and some personal experiences on ignorance and the lack of understanding. I also talked of my personal involvement in research with people from different cultures, religions, customs, and ways of life. Working together on science helps people not only to augment their scientific knowledge, but also to know each other as people.

SUMMARY OF MEHDI BAHADORI'S PRESENTATION

Science alone is not the answer. Science plus morality or ethics can solve problems. We need to have moral values to use science for the benefit of mankind.

Comments

Sobouti: In the ancient Spartan society, if you stole something and you weren't caught, you were a brave man. But if you were caught stealing, you were just a thief. You see, this is the ethic of one primitive society. My point is that ethics and morality depend on the societies and on the cultures.

Bahadori: I'm not going back to five thousand years ago. We are the people of today. Why should you refer to the ethics of five thousand years or two thousand years ago? Today's science and today's definition of ethics are what we should use. If a man is a terrorist, how can we justify his actions? Just two days ago, 160 people were killed and hundreds were injured in Pakistan. Those responsible used the most recent scientific knowledge. Certainly their action is unethical.

Sobouti: I agree with you on this incident. Right now, however, there are people in the world that you and I are calling terrorists and others are calling democracy seekers and freedom fighters and vice versa. Even these words do not have clear-cut definitions.

16

Closing Remarks

NORMAN NEUREITER
American Association for the Advancement of Science

I want to look beyond the "Gateway" and consider the possibilities for turning the many suggestions into an enduring reality. If practiced by scientists of various countries—either bilaterally or multilaterally—scientific cooperation, in addition to providing direct scientific benefits from working together and solving common problems, can also serve as a gateway for better understanding between nations. If there is a shared belief of all the cooperating parties in the validity of the scientific method and the truth that can be derived from the practice of evidence-based science, then that is a head start toward the achievement of mutual understanding.

It is also useful to consider specific mechanisms for cooperation that can bring about such benefits. Participating in a joint seminar or conference or workshop on one or more specific topics is one form of cooperation. One meeting can be useful; but to have lasting impact there should be structured follow-up activities, such as convening a continuing series of workshops or seminars around the same or related topics over an extended period of time. Another possibility is to identify specific scientists or institutions that will continue to work together on problems identified in the seminar or workshop, thus building institutional and personal relationships as well.

Cooperation can take other forms. There can be short visits of senior scientists and professors, for whom workshops or lectures are convened in the receiving country. Two groups in two countries can work on different aspects of the same topic, with easy communication over the Internet and occasional short-term visits to discuss results. A truly cooperative project is best achieved through actual exchanges of researchers between laboratories in the two countries. The exchanges can be at any level academic level—bachelors, masters, postdoctoral, or senior researcher. A successful project should result in joint publication of results in an international journal by authors from both countries. Such joint projects can be undertaken through informal relationships between individual scientists and their coworkers or they can be institutionalized through formal agreements between institutions.

If political relations between two countries are strained, it is sometimes still possible to develop cooperation through international organizations such as United Nations affiliated organizations, the International Council of Scientific Unions, or smaller bodies such as the International Institute of Applied Systems Analysis. Professional scientific societies, such as the American Chemical Society or the academies of science of different various countries can also facilitate cooperation.

There are many areas in which cooperation is possible, even as there may be some topics or areas that will be considered too sensitive for cooperation. Promising areas for cooperation include science education at all levels, science and technology policy, geosciences, energy, food production, water quality and quantity, and medical research problems, together with basic research in many disciplines.

Turning to the example of the United States and Iran, efforts are needed on both sides to eliminate or mitigate the barriers that inhibit cooperation. On our side, we can generate publicity about our present visit to Iran, the success of our seminar, and the warm welcome that we have received. We can mention some of the cooperative research opportunities that exist in Iran. We can also work with the U.S. Department of the Treasury to facilitate the issuance of licenses to work with Iran and with the U.S. Department of State to facilitate the issuance of visas for Iranian scientists. We need private sources of funding for cooperation, and we should explore the interest of Iranian-born scientists now living in the United States in promoting cooperation with Iranian institutions. The success in some of these actions will also depend on the public perception in the United States of Iran's international posture and the public policy positions taken by the Iranian government. In other words, there is a lot for us and our scientific institutions to do in both countries to improve the climate for U.S-Iran cooperation in science and technology.

Today we are at a very difficult moment in U.S.-Iran relations. We need real efforts on the part of scientists in both countries who believe that the correct future is a better relationship between our countries. We need to work hard to make that point clear as we go forward in order to change the present course in our relations. I fear that the present direction is toward a further deterioration that would be most unfortunate for both sides and deny both our peoples the benefits that are possible from a different relationship and closer cooperation.

A

Workshop Agenda

Science as a Gateway to Understanding

TEHRAN, IRAN

October 20, 2007

Opening Session

Words of Welcome
Mohammad Kazem Bojnordi
Center for the Great Islamic Encyclopedia, Iran

Where Does Science Go?
Honorable Mohammad Khatami
Former President of Iran

Scientists and Truth
Reza Davari Ardakani
Academy of Sciences of Iran

The Innovation Ecology
William Wulf
U.S. National Academy of Engineering

Session 1
Chair—Yousef Sobouti

Wisdom, the Best Gateway to Understanding
Mostafa Mohagheh Damad
Academy of Sciences of Iran

InterAcademy Cooperation: An Approach to Understanding
Abulhassan Vafai
Sharif University of Technology, Iran

Session 2
Chairs—Abulhassan Vafai and William Wulf

Understanding Others, the Science Way
Yousef Sobouti
Institute for Advanced Studies in Basic Science, Iran

Loving and Sharing Science: Pierre-Gilles de Gennes
Etienne Guyon
Ecole Normale Supérieure, France

Successes in Building International Bridges through Science
Norman Neureiter
American Association for the Advancement of Science

Session 3
Chairs—Abulhassan Vafai and William Wulf

The Scientific Work of Abu Reyhan Biruni as the Mirror of the World's Science
Hassan Tajbakhsh
Tehran University

Bridge for Mutual Understanding
Akio Matsumoto
Chuo University, Japan

October 21, 2007

Session 4
Chair— Mehdi Bahadori

The Universality of Science: Examples from History
Hossein Masoumi Hamedani
Sharif University of Technology, Iran

The Role of International Scientific Organizations
Michael Clegg
U.S. National Academy of Sciences

Science and Culture
Bernard Maitte
Lille University of Science and Technology, France

Session 5
Chair— Mehdi Bahadori

A Half Century of Successes and Problems in U.S.-Iranian Cooperation in Science, Engineering, and Medicine
Glenn Schweitzer
U.S. National Academy of Sciences

Opportunities for International Collaboration in Earthquake System Science
Thomas Jordan
University of Southern California

Session 6
Chairs—Mostafa Mohagheh Damad and Norman Neureiter

Science, Technology, and Innovation in the Knowledge Economy: Prospects for Cooperation
Ousmane Kane
African Regional Center of Technology, Senegal

Global Energy Challenges
William Colglazier
U.S. National Academy of Sciences

Working with People with Different Cultures and Backgrounds
Ferenc Szidarovszky
University of Arizona

Session 7

Open Forum
Yousef Sobouti

Closing Remarks
Norman Neureiter

B

Workshop Participants

Mehdi Bahadori
Vice President for Research
Academy of Sciences of Iran

Mohammad Kazem Bojnordi
President
Center of the Great Islamic Encyclopedia

Michael Clegg
Foreign Secretary
U.S. National Academy of Sciences
and
Donald Bren Professor of Biological Sciences
University of California, Irvine

William Colglazier
Executive Officer
U.S. National Academy of Sciences

Reza Davari Ardakani
President
Academy of Sciences of Iran

Etienne Guyon
Professor
Ecole Normale Supérieure, Paris
and
University of Paris

Thomas Jordan
Director
Southern California Earthquake Center, University of Southern California

Ousmane Kane
Executive Director
African Regional Center for Technology

Mohammad Khatami
Former President of Iran

Bernard Maitte
Professor
Universite de Lille

Hossein Masoumi Hamedani
Sharif University of Technology
and
Academy of Persian Language and Literature
and
Center for the Great Islamic Encyclopedia

Akio Matsumoto
Professor of Economics
Chuo University

Mostafa Mohagheh Damad
Head of Islamic Studies
Academy of Sciences of Iran

Norman P. Neureiter
American Association for the Advancement of Science

Glenn Schweitzer
Program Director
U.S. National Academy of Sciences

Yousef Sobouti
Director
Institute for Advanced Studies in Basic Science, Zanjan
and
Academy of Sciences of Iran

Ferenc Szidarovszky
Professor of Systems and Industrial Engineering
University of Arizona

Hassan Tajbakhsh
Professor
Tehran University
and
Academy of Sciences of Iran

Abulhassan Vafai
Professor
Sharif University of Technology

William Wulf
President Emeritus
U.S. National Academy of Engineering
and
University of Virginia